BALL OF
Confusion

JOHNNY BALL

BALL OF

Confusion

Puzzles, Problems & Perplexing Posers

ICON BOOKS

Published in the UK in 2011 by
Icon Books Ltd, Omnibus Business Centre,
39–41 North Road, London N7 9DP
email: info@iconbooks.co.uk
www.iconbooks.co.uk

Sold in the UK, Europe, South Africa and Asia
by Faber & Faber Ltd, Bloomsbury House,
74–77 Great Russell Street,
London WC1B 3DA or their agents

Distributed in the UK, Europe, South Africa and Asia
by TBS Ltd, TBS Distribution Centre, Colchester Road,
Frating Green, Colchester CO7 7DW

Published in Australia in 2011 by Allen & Unwin Pty Ltd,
PO Box 8500, 83 Alexander Street,
Crows Nest, NSW 2065

Distributed in Canada by Penguin Books Canada,
90 Eglinton Avenue East, Suite 700,
Toronto, Ontario M4P 2YE

ISBN: 978-184831-348-4

Typeset in Palatino by Marie Doherty
Printed and bound in the UK by
CPI Group (UK) Ltd, Croydon, CR0 4YY

For Four Grandchildren
Woody, Ronnie, Nelly May and Albie

Just like your Grandad – life is always a bit of a puzzle!

Contents

Foreword

If I had a pound for every time someone had told me
how my Dad's television shows had made science
and maths a joy for them when they were a kid,
which in turn had encouraged them to follow a career
in engineering or medicine or some other clever
vocation … well then, I'd have a ridiculously enormous
pot full of coins!

Exactly how many coins would be in that pot would
take some figuring out – it might involve some guess
work, perhaps some logical thinking. Of course you
could just empty the pot and count the coins but that
would be far too obvious. At this point my Dad would
wade in and give a fantabulous explanation of averages
and probability.

With my heritage you'd think I'd be good with numbers,
a natural. Sadly this is not the case. The thirst for
scientific knowledge and understanding of all things
numerical is, apparently, not handed down in the genes.
One might even argue that when it came to me the
intelligence gene skipped a generation. I was an average
maths student. I was only really good with digits when
it involved remembering boys' telephone numbers. 'But
you have the best teacher in the world right there in
your Dad!' people would say. Yes, occasionally I would
ask for some help with my algebra homework and
my Dad would ooze enthused explanations about the
Egyptians' weighing systems and Pythagoras' theorem

which, although fascinating, still didn't explain to *me* why $x^2 + y^2 = z^2$.

It's true, my Dad knows an awful lot about a huge number of things. What he doesn't know about the history of science and mathematics is probably not worth knowing. His mind is like an Escher drawing; a never-ending maze of facts and figures about everything from Archimedes, Roman road building, gravity and trigonometry to Johannes Kepler, pi, space exploration, the universe … I could go on.

The best bit is that it's all self-taught – he went to grammar school but never to university. He just read books – great books. He's written a few good ones himself. He loves learning about how things work, how men and women have made such invaluable discoveries and how there is maths behind everything around us. Dad believes maths and sciences are terrific, important subjects and enormously fun to learn if taught with enthusiasm and energy.

Family gatherings in our house can be great fun – heated debates about who were better, the Egyptians or the Romans, arguments about global warming or over which is the most valuable number. The grandkids in the family quite rightly think Granddad knows everything – a taxi driver recently joked with me, 'Most people have Google, your kids have Granddad.'

I'm so proud of my Dad and all that he's done for the sciences and for education. He's a very clever, brilliant chap, so it's an absolute pleasure to share him with

everyone. This especially includes the people who listen
to my radio show, where we do a weekly feature called
Ball of Confusion in which Dad sets puzzles. I, of course,
rarely get them right but when I do it's such a thrill – I
even surprise myself. I do hope you and your family
enjoy this collection of some of the best.

Zoe Ball
Summer 2011

How *Ball of Confusion* was born ...

When my daughter Zoe began the BBC Radio 2 Saturday Morning Breakfast Show in 2008, her producer asked for ideas for special features. Zoe asked me if I could do a puzzle of some kind on each show. I would set the puzzle in the first hour and give the answer and explanation an hour or so later.

At first, I was worried that I wouldn't be able to find enough puzzles that would work on radio, but I soon found that there was really no problem at all. Radio is such a wonderful medium, in that you can paint an image or set a scene in just a few words.

Zoe and I have now delivered well over 100 puzzles together and I am amazed to find that I am still nowhere near to running out of ideas. I have all my puzzle experience, which began at school, as well as my many old books on all things mathematical to thank for that.

I should mention that on the show, Zoe very seldom comes up with the answer to a puzzle. Now, that could be because she is really quite stupid at puzzles, but it isn't that way at all.

When the show is transmitted the listener has an hour to deliberate and try to solve the problem before we give the answer. But for Zoe, there is no thinking time at all. For one thing, she has the rest of the show to do!

But more importantly, the puzzles are all pre-recorded. Otherwise I would be making a trip into the London studios every Saturday morning before dawn, for five minutes work. So about three times a year, we record a whole bunch of puzzles and for each, I deliver the question, and then almost instantly I launch into the answer and explanation. So Zoe really is allowed no thinking time at all.

She is actually quite good at puzzles – where she gets it from, I don't know. Just like me, though, she is often slow to show confidence. However, given time and encouragement, she is pretty good at everything she sets her mind to. But I would say that wouldn't I? I'm her Dad.

Hey, that's an idea for my next book – 'Ball of Compliments!'

Johnny Ball
Summer 2011

Introduction

When I was a comedian, it soon became apparent that there was no such thing as an original joke. Whatever the wise-crack, pun, witticism, tall tale, spoonerism, one-liner, play on words, rudeism or gag, it could always be proved to be a newer version of something that had been said before.

So it is with puzzles. If you want an early comic puzzle, try this! Why, in an Egyptian tomb did someone long ago inscribe the words (or rather hieroglyphs) that said, 'Can I borrow your washing line? Someone's spread jam on mine!' It is clearly a joke, but the puzzle is who thought of it and chose to record it for posterity?

Almost all the puzzles in this book, though re-set or juggled about by me, were created by people in the past who revelled in puzzles. As with any good joke, there are always people wanting to pass puzzles on for others to enjoy.

Unlike most puzzle books, you will often find my solutions longer and more multi-faceted than the basic answer. This is because puzzles are to be understood, solved and learned from – that is, and always was, their purpose. To set a question and then, on a later page, give the answer as a single word or number would be of no use in helping a puzzled puzzler, who did not understand the question in the first place, to understand how that answer was arrived at.

So I have indulged myself and my love of puzzles
by trying, in most cases, to extend the basic idea
with mathematical thoughts that take the puzzler
a step further and into a deeper realm of puzzle
understanding. On occasion, as with the jug pouring
puzzles, which I love, I will say, 'If you liked that
puzzle, then here are a few more in the same vein.'
Why? Because when a basic understanding dawns
on a puzzler, they need to extend that understanding,
to broaden their knowledge and appreciation of the
variations on each particular theme. This is why people
get hooked on Sudoku.

Sometimes I will credit the originator of a puzzle,
because from my researches, I happen to know just who
that person was. But on many occasions there will be
no credit, even though I may have first encountered a
puzzle in a book written by someone with a knowledge
of puzzles far greater than my own. The reason I feel I
can repeat the puzzle without giving them credit is that
I am pretty certain they *also* found it written by someone
who had gone before *them*.

In my modern day maths lectures I often open with a
slide showing the Vitruvian Man, as drawn by Leonardo
da Vinci. It shows a man set in a square with his arms
outstretched, demonstrating that a man's height is
almost always the same as the distance his fingers can
reach sideways. The Romans called both distances a
man's 'stature'. The Vitruvian Man is also set in a circle,
with a second set of arms reaching higher, to the very
edge of that circle.

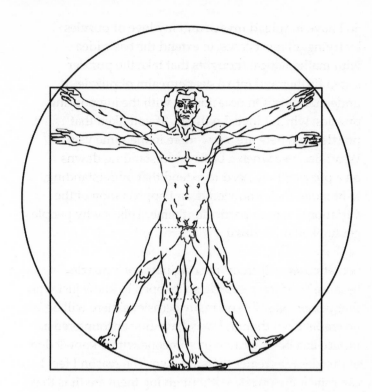

Have a look at the picture, then cover it and ask
yourself, 'What point is half way up the human body?'
Well? In my lectures, I invariably find that almost
everyone will say, 'Your waist, belly button or navel!'

But they are all totally wrong. Half way up your body
is your pubic bone, at the point near the bottom of
your torso, just above where your body starts to get
interesting. The navel or belly button is about a span
higher – the distance between little finger and thumb
when a hand is outstretched.

Look at the Vitruvian Man and you will see that the navel is at the centre of the circle, not the square. This point is one span higher than half way up your body. If you reach either arm straight up above your head you will reach a point one cubit or two spans higher than your height. So the Vitruvian square is eight spans high by eight spans wide, but the circle has a radius of five spans and a diameter of ten.

But by far the most important thing about Leonardo's Vitruvian Man is that from it we can understand exactly what learning is all about. Where did Leonardo, this man of a genius hardly even equalled in the past 2000 years, get the idea for the drawing? Simple – he nicked it from the architect Vitruvius, who designed the Roman Colosseum 1500 years before Leonardo was born. Sadly, the Colosseum itself eventually went broke – the lions ate all the prophets.

The tale of Leonardo and Vitruvius explains that all learning is theft. Almost all top sportsmen and women talk of the idols that inspired them when they were young. They learned to copy their idols until eventually they became just as good, if not better – it is the natural way the whole of mankind has developed and improved over time.

A child should never be allowed to be worried or doubtful about their ability to learn. In their first two years of life, they mastered an incredibly complex language and learned to mimic every aspect of the everyday antics, accents and actions of their parents.

So, it is in our nature to steal the ideas of those who have gone before and build upon them.

This book is full of stolen ideas, as I know that those who taught me learned in the same way I did, and learning is arguably the only thing absolutely worthwhile in all our lives, including our learning how to love and be loved.

I love people and I love puzzles. Hopefully I will be able to blend those two loves in the following pages by helping you, dear reader, to love puzzles and puzzling and to love learning and acquiring knowledge. Doing so will broaden your mind, your understanding, and eventually your whole outlook and personality.

How bright do you need to be to enjoy puzzles? Well, let me tell you a little about my own education. I was born in Bristol to parents who loved games which involved counting and calculating, so I was naturally very good in primary school and always top in maths, passing my '11 plus' with ease.

About that time my parents moved to Bolton in Lancashire – but I found them and I moved too. I was accepted into Bolton County Grammar School and to reflect my success at primary school, I was placed in form 2B for the first year. Then in subsequent years, they placed me in form 3C, then in 4D, then in Lower 5E and finally in 5E, because they didn't have a 5F. I had always been top in maths but I left school a failure, with O Levels in just two subjects, one of course was maths – the other one wasn't.

So I joined an illustrious list of school failures including Winston Churchill, Richard Branson and a German lad whose teacher declared he would 'never make anything of himself'. Luckily his engineer uncle showed him the maths puzzles that I was to discover 60 years later, many of which feature in this book.

This activity trained and readied the lad's mind for the path ahead, where in just one year – 1904 – he produced three scientific papers explaining to the world for the very first time:

1. How energy creates all light;
2. That molecules constantly crash into each other in a liquid;
3. His very Special Theory of Relativity.

This puzzle-loving lad became the greatest scientist of all time – Albert Einstein.

So, becoming a puzzle lover might place you in very illustrious company indeed.

But the aim of this book of puzzles, old and new, is simply to help you discover the fun of exercising that brain of yours, which makes you the totally unique individual that you are. Enjoy.

Johnny Ball
Summer 2011

1

Kitchen Capers and Domestic Problems

Many of the puzzles in this book are versions of puzzles which were thought up and amusing people long before TV and even radio were invented. In those days, people were nowhere near as worldly as we are today. Many people never strayed far from their home town or village in their entire lives.

People in those days made their own entertainment from the things that surrounded them in everyday life and the things that they thought they knew about. Just like today, much of their fun and humour was all about playing tricks on their family and those around them.

So, in this chapter the puzzles are all about things that even today, you might link to everyday home and social life.

1. Bun fun with Mum

As an Easter treat Mum has made 27 hot cross buns. As a surprise, she put a £2 coin inside one of them. Unfortunately, she can't remember which one – except that it is clearly heavier than the rest. The odd bun is not heavy enough to detect the difference by hand, though!

You have a simple set of balance scales, but no weights. Can you find which is the heaviest bun (and so snaffle the £2) in just three weighs?

Answer on page 117

2. The sands of time

You have a couple of glass sand timers. One takes 7 minutes for the sand to trickle through and the other takes 11 minutes for the sand to trickle through. Using these two, can you time a 15-minute passage of time, exactly?

Answer on page 117

3. One green bottle

Bottles usually have a circular base and a tapered neck.

You can work out the cross sectional area, as it is πr^2.

If, for example, the wine bottle is 7 cm across, its internal width will be about 6 cm across. The internal radius is 6 ÷ 2 = 3. Its cross sectional area is $3 \times 3 \times 22 \div 7 = 28.28$ cm².

But for simplicity, let's say we know a bottle's cross sectional area to be 30 cm².

If it were a cylinder, with no tapered neck, the volume of the bottle would be easy to find: height × 30 cm².

But it does have a tapered neck. It also has some wine in it. How can you work out the volume of the bottle?

Answer on page 118

4. A toast to toast

Isn't making toast a bind if you haven't got a toaster? You only have a grill and space for 2 pieces of bread under it. So it takes 4 toastings to do 3 slices on both sides. Or does it?

Can you toast 3 pieces of bread on both sides in less than 4 toastings?

Answer on page 119

5. Time on his hands

Let's go back to the days before we had telephones and mobiles phones – oh bliss! A chap forgot to wind his only clock and it stopped. He knew his pal had a clock that always told the right time. So, he walked to his mate's house, where he stopped and had a chat and a cup of tea. He then walked back home, went straight in and set his clock to the right time. But how did he manage to do this?

Answer on page 119

6. Draw your own conclusions

In the playroom, I have just put an elastic band around a bundle of coloured pencils. The bundle now forms a perfect hexagon. Counting around the outside of the bundle, there are 18 pencils. How many pencils are there in the bundle?

Answer on page 120

7. Sock it and see

I have 3 pairs of socks. I actually have a few more, but that has nothing to do with the puzzle. I have a blue pair, a green pair and a red pair. I want to hang them on a washing line, but in a puzzling mathematical way. I want to hang them so that:

> There is 1 sock between the blues
> There are 2 socks between the greens
> There are 3 socks between the reds.

In what order along the washing line should I place them?

Tip: start with the biggest numbers!

Answer on page 121

8. Put a sock in it

Here is a similar puzzle but with 4 pairs of socks.

You have 4 pairs of socks, blue, green, red and yellow. Hang them on the line so that:

> There is 1 sock between the blues
> There are 2 socks between the greens
> There are 3 socks between the reds
> And there are 4 socks between the yellows.

Answer on page 121

9. A sucker born every minute

You have 5 sweets and you suck one every 10 minutes. How long will it be before you have none left?

Answer on page 122

10. An eggs-acting question

Mary sets up her egg stall at the side of a country road and waits for her customers.

Anne arrives first and buys half the entire egg stock, plus half an egg.

Betty turns up and buys half the remaining eggs, plus half an egg.

Celia then buys half the eggs that are left, plus half an egg.

Now Daisy arrives and, while Mary isn't looking, steals half the eggs and half an egg.

Mary looks round to find she has no eggs left at all! Mary leaves her egg stall empty handed and eggsasperated.

Not one egg was broken. So, how many eggs did Mary start with?

Answer on page 122

11. Put your money where your mouth is

Here are 6 identical touching coins.

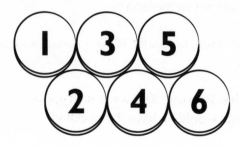

Can you, in just 3 moves, change their position so that they form a hexagonal circle around a single coin sized space? It's easy in 4 moves, but can you do it in 3? The secret is in finding the first correct move.

Answer on page 123

12. Fumbling in the dark

The lights have gone out – I said we should never trust wind farms – and I have to search my wardrobe in the dark for a pair of shoes and socks.

I have 3 pairs of shoes, 12 pairs of black and 12 pairs of brown socks.

How many of each do I need to take, to be sure I have a matching pair of each to wear?

Answer on page 124

13. Animal farm

I was once looking for work and I met a farmer. I said, 'Can you use me on the land?'

He replied, 'Well, not really. We use special stuff for that.'

A farmer has 20 goats, 30 cows and 50 horses. How many horses would he have if you called the cows 'horses'?

Answer on page 124

14. The weight of the matter

Corporal Jones in *Dad's Army* was a butcher and he had a pair of weighing scales. The set of weights in those days comprised of one each for 1 ounce, for 2, 4, 8, 16 ounces (or 1 pound) and for 2 and 4 pounds. With the seven weights he could weigh up to 127 ounces, or 7 pounds 15 ounces. Then we went metric to simplify it all and sure enough everything got more complicated.

But even today, with a set of balance scales it is still possible to weigh every weight from 1 unit to 121 units, using just five different weights. What are the five unit weights? Also, were there to be six weights in the set, what would the next one be?

Answer on page 125

15. You could lose your balance

This puzzle first appeared in 1914 in Sam Lloyd's Cyclopedia of Puzzles.

You have a set of balance scales and some bottles, glasses, plates and jugs. There are three sets of items that balance perfectly.

(1) A bottle and a glass balance a jug
(2) A bottle balances a glass and a plate
(3) Two jugs balance 3 plates.

So, how many glasses would exactly balance a bottle?

Answer on page 125

16. A pond to ponder over

Last year a man bought a water lily plant for his garden pond. He was told that from the day he planted it, it would double its size every day. After 30 days the water lily plant completely covered his pond.

Unfortunately it died over the winter. However, this year the man has bought two water lily plants. How many days will they take to cover his pond?

Answer on page 126

17. Early doors

Think of your house number (or just imagine one). Keep it to two digits. Double it. Add 5. Multiply by 50. Add your age. Add the number of days in the year, 365. Now subtract 615.

Check the result. The first two digits are your house number, the last two are your age. But why?

Answer on page 126

18. Measure for measure

When I was a toddler, my granny's milk came from a horse and cart. No, I don't mean it was horse's milk. It was cow's milk, delivered by horse and cart. The horse knew exactly when and where to stop and even how long to wait before he walked on.

The milk came in a large milk churn and the milkman using a half pint, pint or a two pint measuring can and he would ladle it into the customers' milk jugs. Granny and all the other customers had their own milk jugs (no need to recycle bottles in those days), plus a square piece of gauze with beads at the corners to cover the milk and keep the birds from stealing the cream.

Now, here is a fine but very old measuring puzzle, that also leads to more of the same kind.

You have three milk jugs which will hold 8, 5 and 3 units of milk. The 8-unit jug is full. Can you divide the milk into two 4 unit measures, and not spill any milk? There are two ways of finding the solution.

As this is a well-known puzzle, here is an additional puzzle. With these jugs, It is possible to measure every unit amount – 1, 2, 3, 4, 5, 6 or 7. In finding one solution you will achieve all unit measurements except one. In working through the other solution you will find them all except a *different* one. Which solution misses which unit measurement on the way to that solution?

Answer on page 127

19. Milk of human kindness

Here is another milk jug problem for you to 'pour' over.

Many years ago a milkman arrived at two cottages at the end of a lonely lane. He had two full 10-litre churns of milk left. Two ladies came out of their cottage doors to greet him. One held out a 5-litre jug, the other a 4-litre jug, the only jugs that they possessed.

Then the problem started. They both asked for 2 litres of milk. He could have filled the 4-litre jug and told them to split it between them. But their two jugs were of different shapes, so there would be no way of knowing when each had exactly 2 litres.

Still, by some clever pouring the milkman managed to deliver exactly 2 litres to each lady. Then he went on his way, leaving both his customers happy. Can you work out how he did it?

Answer on page 128

20. Going for the juggler

There are many 'two jugs and a tap' problems. In this one you have a 7-pint jug, a 10-pint jug and a tap. The task is to measure out exactly 9 pints of water. Oh, and each time you fill the second jug, you must empty it down the sink. It will waste a lot of water, but it is the only way to arrive at the solution. Have a go. First question is which jug will you fill first?

Answer on page 129

21. More jug juggling

Some people love these jug and tap problems. If you are one of them, here are three different puzzles for good measure. For each one there are two solutions depending on which jug you fill first. I do hope you choose right – one solution sometimes takes considerably more moves than the other!

Measure exactly 1 unit of water, starting with the following pairs of jugs:

  A 7- and a 9-unit jug
  A 7- and a 12-unit jug
  An 8- and a 13-unit jug.

Good luck!

Answer on page 130

22. Firing on two cylinders

This puzzle is quite different from all the previous jug puzzles.

You have two perfectly cylindrical jugs. One holds 8 pints and is full of milk (or water). The other holds 5 pints but is empty.

Can you produce exactly 4 pints in each jug? Remember this time you have no 3-pint jug. And, most important of all, can you do it in just one move?

Answer on page 131

2

'I Can Hear You Thinking' Puzzles

This chapter features puzzles that will hopefully make you think for some time. The inspiration for the title came from my wife, who has often said in bed at night, 'I can't sleep for listening to you thinking!'

In Rudyard Kipling's famous poem *If* there is the line, 'If you can think, yet not make thoughts your aim.' Thinking requires an aim or a purpose beyond itself – so tackling puzzles with the aim to finding a solution must be a worthwhile exercise.

Some of these puzzles are easy and some are quite tough, but the answers will not come unless you give them each some 'thinking time'.

So, read the question carefully, then jot down a few ideas and see if you can puzzle out the answer – for that is what puzzles are for. No looking up the answer unless you really have to! Enjoy.

1. On the right lines?

I once had a job as a shunter. Not on the railways – in a department store. I would look out for someone stealing something and say to them, 'You shunter done that!'

Anyway, this next puzzle *is* to do with railways. Here is a map of a triangle of railway lines.

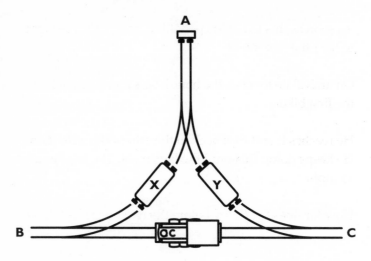

There is a locomotive engine and two goods vans. The goods vans are labelled X and Y. The lines at B and C connect to the main lines, but the dead end line at A can only take either of the goods vans – the locomotive cannot fit into the space.

Can you get the locomotive to swap the position of goods vans X and Y?

Answer on page 132

2. Between flights

Two lads live 20 miles apart and agree to cycle to meet each other.

They both cycle at 10 mph.

A fly, having nothing better to do, is sitting on the handlebars of one lad's bike.

As soon as this lad starts pedalling, the fly flies straight to the other lad's bike.

On arrival he touches the handlebars and flies back to the first bike.

He reaches it and flies back to the other bike once more. He keeps doing this until the lads meet. The fly flies at 15 mph.

How far does the fly fly before the two boys meet?

Answer on page 132

3. It's a knockout

If you have 8 teams in a knockout, you play 4 matches and reduce it to 4, then 2 matches and reduce it to 2 and then the final will leave just 1 winner. If you have 29 teams, how many matches will you need to find a winner? The answer is easier than you might at first think.

Answer on page 133

4. The Inn of the Sixth Happiness?

Six fellas arrive at a hotel for a stag weekend which they had booked ahead. Unfortunately, on arrival they discover that there are only 5 rooms available.

'Don't worry,' says the hotel manager, 'I can fix this. I'll put the 1st lad and the 6th lad in room 1, just for the moment. Now I'll put the 3rd lad in room 2 and the 4th lad in room 3. Next I'll put the 5th lad in room 4. Now I can take the extra lad, the 6th, from room 1 and put him in room 5. Now everybody is happy!'

Now, what is wrong with all that?

Answer on page 133

5. Talking balloons

A fella picked up his son from a party, in his wife's new car. Little Johnny had a helium balloon, on which he had just painted a portrait of Mummy, in wet yellow paint.

So, he got Johnny into the car and drove very carefully, so that the balloon stayed in the middle of the car, straight up on its short string, so as not to get paint on the upholstery.

Just as they got home he turned LEFT into his drive – but had forgotten about the balloon which swung wildly sideways and got paint all over the interior of the car.

But on which side of the car?

Answer on page 133

6. Between the sheets

A newspaper has a sheet of 4 pages missing. I can tell you that pages 6 and 19 are missing. It's a simple thing to say which other pages are missing. But can you say how many pages the original newspaper had altogether?

Answer on page 133

7. Old before his time?

A chap once said to me, 'The day before yesterday I was 19 and next year I'll be 22.' On what day was he born and when did he make the statement?

Answer on page 134

8. Flights of fancy

My uncle was an airline executive. He worked at a garage and when your tyre was flat he would get his air line and pump it up again. He knew a lot about inflation, but here's a puzzle.

A girl sat crying at an airport. A passing pilot asked what the trouble was. She said that her purse and air ticket had been stolen and that she couldn't get home.

'Don't worry,' said the pilot, 'I'll drop you off.'

'But you don't know where I'm going!' she wailed.

'S'okay,' he said, 'I can pass that way without going out of my way.'

The question is, where was he going?

Answer on page 134

9. What to wear where

These days, teenage girls sometimes wear less material on a Friday night out than my mum used to keep the flies out of the milk jug.

Three girls turned up to a party in a black dress, a green dress and a white dress. They were Miss Black, Miss Green and Miss White.

Miss Black said, 'It's amazing – our dresses match our names, but none of us is actually wearing the dress that matches her name.'

And the girl in green said 'So what?'

So, what I want to know is, who was wearing which dress?

Answer on page 134

10. The floating hat

When I got on my mum's nerves, by just being a kid, she would often say, 'Why don't you go for a paddle in the river till your hat floats?' Her mother had probably used the same expression to her. Here's the puzzle …

Flow at 4mph

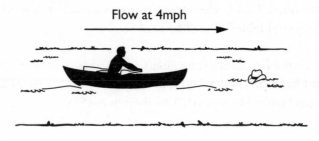

A rower decides to go for a row upstream on a lovely summer's day. The river is flowing at a steady 4 mph. As he gets started he doesn't notice that his hat falls into the water. It starts to float down stream. He rows upstream for 3 minutes at a steady 8 mph and then notices his hat is missing. He immediately turns around and rows back at the same rowing speed.

How long will it take him to reach his hat?

Answer on page 135

11. Truth and Liars Club

When I first read P.G. Woodhouse's Bertie Wooster Stories, I was an immediate fan. This puzzle involves a fictitious gentleman's club, similar to Bertie's Drones Club.

Every member of the Truth and Liars Club is either a solid 100% truth-teller or a cast iron perpetual liar.

One evening, a member of the club rang me at home and said, 'We've just had dinner on a round table and each man said, "I declare, the man on my left is a liar."'

Then the phone went dead. I rang back but someone else answered. I said, 'How many men were seated around the round table?' The chap said 11.

Then the first caller rang again and said he was sorry we had been cut off. I asked him how many men were at the round table. He replied, 'Oh, there were 8.'

Why, I wondered, did I get two different answers? I realised that one of the callers told the truth and one told lies. But which one?

Answer on page 135

12. Smitten by kittens

The lady living next door to me loves cats. If a stray appears near her house she offers it food and soon it has become another of her cat and kitten clan. The other day I asked her how many cats she had at present and she said, 'Oh, not many right now – in fact I have ¾ of their number plus ¾ of a cat.'

How many cats does she have?

Answer on page 136

13. Suitably attired

Benny Hill used to talk about a new synthetic cream made out of wood, coal and rubber. You didn't get indigestion. You either got a splinter, a clinker or a blow out!

A cyclist is on a journey and after completing ⅔ of the distance, he has a blow out – a flat tyre. He walks the rest of the way and spends twice as long walking as he did riding. How much faster than his walking speed, did he cycle?

Answer on page 136

14. On political lines

Remember the peer who dreamed he was making a speech in the House of Lords? Woke up and found that he was.

An MP on a train fell asleep half way to his destination. He slept until there was half as far to go as he had gone while sleep. For how much of the journey was he asleep?

Answer on page 137

15. Brief lives

Twin boys were born together, discounting a slight pause by Mum between delivering each of them. They lived their lives travelling the world, but eventually they died, coincidentally also at the same time. But when they died one was older than the other. Can you explain how that could be?

Answer on page 137

16. Merrily we roll along

I once knew a girl who was told by the doctor that she had Egyptian flu. Sure enough, a few months later she became a mummy.

Ancient Egypt has always been a puzzle. For example, how did the ancient Egyptians move the heavy stones with which they constructed their pyramids? Did they use logs as rollers?

Actually, most tomb paintings suggest that they used logs as sled runners and that, with enough manpower, any weight could be shifted this way. Fine sand might even have been used as a lubricant under the huge runners.

I know of no tomb painting that shows them using logs as rollers – they would have been impractical because as the load moved forward, the logs would keep coming out at the back. Everything would have to stop while the rollers were manhandled to the front again. Also, if any two rollers touched they would bind on each other and everything would stop dead.

Nevertheless, here is a puzzle about moving a slab using rollers.

You have a heavy stone slab on four identical rollers.

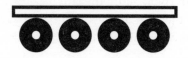

Each is 1 metre in diameter and 3½ metres in circumference.

As each roller makes one complete turn, how far will the slab move?

Answer on page 137

17. It's in the bag

Here is one of Lewis Carroll's favourite puzzles.

A bag contains a counter. It is either black or white, but you don't know which. You now put a white counter in and you shake it all about. I was always hopeless at the Hokey Cokey – I used to be putting it in when everyone else was shaking it all about. Sorry, I digress – you shake the bag and take out one counter. Wow, it is white!

Now, what are the chances of the other counter being white?

Answer on page 138

18. The vampire umpire

The Transylvanian tourist guide has a problem. He needs to get two tourists and himself across a Transylvanian river. Sadly the sun has just gone down and what do they see on the other bank? Three lusty young vampires wishing to cross the other way.

There is a canoe that takes two people. However, if ever the guide or the tourists are outnumbered by vampires on either bank, it is slurp-slurp, gurgle-gurgle time.

Can they all cross safely without anyone having their neck punctured? The boat must of course start on the vampires' side.

Answer on page 139

19. The prisoner's get-out clause

A prisoner is languishing in a cell in a very unusual jail. There are just 9 cells and only 1 exit door on the outside wall. But there is a door in every internal wall, linking all the cells.

One day, the jailer says, 'Today is my birthday and because I'm in a good mood I'm leaving all the doors unlocked. You can leave the jail and make your escape, but only if you enter each and every cell once and once only on your way out.'

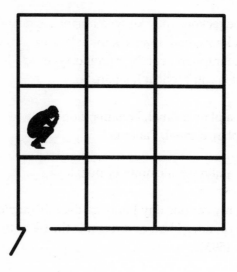

Can you help the prisoner escape?

Answer on page 139

20. Backward drawbacks

Kids often love palindromes, which are words or statements that read exactly the same forwards or backwards. Of course they have to make sense. Here are a few famous ones:

Perhaps the first ever palindrome occurred in the Garden of Eden, with, 'Madam, I'm Adam.'

And Napoleon may well have told Josephine, 'Able was I ere I saw Elba.'

If anyone reckons they know a longer palindrome than any you know, here is a trick to make sure you win that bet. Simply repeat their palindrome, add in, 'sides reversed is', and repeat it again. So:

'A man, a plan, a canal, Panama, sides reversed is, a man, a plan, a canal, Panama.'

Here's a palindromic number puzzle.

Driving my car one day I noticed the mileometer. The number of miles I had covered was the palindromic number 15951.

'Well,' I thought, 'that won't happen again for a long time.' But exactly 2 hours later, the mileometer showed another palindromic number. How fast was I going?

Answer on page 140

21. The answer to four across?

A husband and wife each weigh exactly the same, while their two kids each weigh exactly half of that amount – what a neat family. They have to cross a river and the boat they have will only take the weight of one adult, or two kids. How do they get across?

Answer on page 140

22. Journey's end

A new city light railway has 15 stations. How many different single tickets will need to be printed so that anyone can buy a ticket from any station to any other station?

Answer on page 141

23. The arrow's flight

A Red Arrows pilot was showing his son a picture of the 9 Red Arrow jets in their famous arrow or chevron formation, with aircraft 5 at the centre.

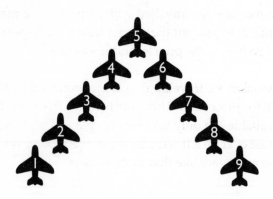

BALL OF CONFUSION

'Which jet were you flying in, Dad?' asked the lad.

'Well,' said Dad, 'if you multiply the number of aircraft to my left by the number of aircraft to my right, you will get 3 less than you would have got if my plane had been 3 to my right.'

It's quite clear the lad's dad was a bit of a maths bore despite being a great pilot. But which jet was he flying in?

Answer on page 141

24. Soul sisters

Tartaglia is the Italian word for 'stutterer'. It is also the name of an Italian mathematician who was born in 1499. When a young lad, his face was slashed by a French soldier and he stuttered thereafter for the rest of his life. His brain, however, certainly did not stutter and he was an accomplished mathematician and a great puzzler.

In those uncertain times no women travelling alone was safe, so they always travelled with a brother or a male body guard, to prevent them from losing their spending money. So here is the puzzle, and it's a cracker.

Three women are travelling, each accompanied by their brother, for protection. They must cross a river, but the boat available can only take two people at a time. No man must ever be left with an unguarded sister. Why? Because men were like that in those days.

Can you get all six people across the river intact? If so, in how many crossings?

Answer on page 142

25. Driven to distraction

Two drivers each have to deliver a package to a city 140 miles away and then come back. One driver travels at 60 mph on the way and 80 mph coming home. The other averages 70 mph both ways.

Who takes the shortest time, or are the two times equal?

Answer on page 143

26. Puffed out

Giving up smoking is not a problem for Fred – he's done it hundreds of times! But this time it is serious. His wife has announced she will divorce him if he has not given up completely in 2 weeks time. He looks into his last packet and there are 9 cigarettes left.

So, how can crafty Fred make his fags last almost, but not quite, 14 days?

Answer on page 143

27. Double expresso not so fasto

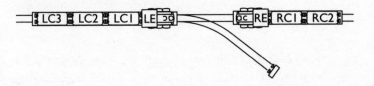

There are two trains on a single line, one has three carriages attached to it and the other has two. The line has a siding which holds only a single locomotive or carriage at a time.

Can you get the two trains to pass and continue on their way, each with the correct carriages?

Answer on page 144

3

Any Number of Puzzles
– About Numbers

Lewis Carroll, whose real name was Charles Lutwidge
Dodgson, taught maths at Oxford University and
produced many new mathematical ideas during his
career. But he always loved to play with maths and
words in a childlike if not childish way. Me too!

This is my adaptation of a quote from *Through the
Looking Glass*. The Red Queen asks Alice, 'If you are
good at sums, then what is [said very quickly] 1 and 1
and 1 and 1, and 1 and 1 and 1 and 1, and 1 and 1 and 1
and 1, and 1?

Alice replies, 'I lost count.'

But if you try it with people and phrase it so that the
1s come in sets of four, it is surprising how many
people, even quite young children, get it right first time,
especially those who are into music. It is the rhythm that
helps people them to count. So:

1 and 1 and 1 and 1, and 1 and 1 and 1 and 1, and 1 and
1 and 1 and 1, and 1 = 13.

1. Puzzle this ONE out ...

Look at this ... $1 \times 1 = 1$
But ... $11 \times 11 = 121$
And ... $111 \times 111 = 12321$

So what does $111,111 \times 111,111$ equal?

Answer on page 144

2. Eight missing

Here is a very special number, with the single digits in order: 12,345,679. Notice that the number 8 is missing. What do you think the answer will be if you multiply it by 9? Try it!

The answer is 111,111,111 (9 consecutive ones).

So, can you guess or work out what $12,345,679 \times 18$ will equal?

Or what $12,345,679 \times 72$ will equal?

Or what the answer will be if you multiply 12345679 by any other two-digit number that is divisible by 9?

Answer on page 145

3. Looking both ways

Take the number 123,456,789 and the number 987,654,321. If you add them together and then add 1, what do you get?

So 123,456,789 + 123,456,789 + 987,654,321 + 987,654,321
+ 2 = ?

Could you write down the answer very quickly?

Answer on page 145

4. Sum for simpletons

Look at this: $12 = 3 \times 4$.

Can you think of four other consecutive digits that do that?

Answer on page 145

5. In seventh heaven

There are 3 six-digit numbers where, if you take the first 2 digits and place them at the rear end of the number, you get the original six-digit number multiplied by 2.

Can you find the three numbers and their doubles?

This sounds like a very difficult problem, but if I tell you that the smallest of these numbers, when multiplied by 7, gives 999,999 all you need is a calculator to find the magic numbers that give the answer.

So, can you find the 3 six-digit numbers and their doubles, all from this one number?

Answer on page 145

6. Pure Gauss work

In 2000 I wrote an educational maths musical called *Tales of Maths and Legends*. This song featured very early in the show:

> *Carl Freidrich Gauss when only nine, he did a sum in record time,*
> *'Please do this sum,' his teacher said, 'add the numbers from one up to a hundred.'*

So, can you 'quickly' tell me the total of all the numbers from 1 to 100?

It would take ages to add $1 + 2 + 3 + 4$ etc., so there must be a simple trick. More importantly, the numbers in the question help lead to the answer! Can you find it?

Answer on page 146

7. Premier division

We've had a few puzzles involving addition and multiplication. Let's try some division.

What is the smallest number that can be divided by all the digits from 1 to 9 without leaving a remainder? That's a number that you can divide by 9 and leave no remainder, by 8 and no remainder, by 7, by 6, by 5 and so on. Try it. It's easier than you think, especially with a calculator.

Answer on page 148

8. An imaginary menagerie

The Wheresitatagen Zoo has many animals, including camels and ostriches.

But how many camels and ostriches does it have?

Well, I can tell you that all together there are 60 eyes and 86 feet. So can you work out how many camels and ostriches there are in the zoo?

Answer on page 148

9. Starter for ten

I once ran the 100 metres in 8.4 seconds – I knew a short cut.

Alf and Beryl have a 100-metre race. Alf wins and as he finishes Beryl is 10 metres behind. When they get their breath back they decide on another race, only this time Alf stands 10 metres back from the starting line. Now who wins this time? Or will it be a draw?

Answer on page 149

10. A Greek bearing birthday gifts

A not so very well-known Greek Philosopher called Sausagyknees was born in 35 BC. If he died on his birthday in 35 AD, how old was he when he died?

Answer on page 149

11. Withdrawal symptoms

I went to my bank the other day but couldn't see Tonto anywhere, which was a surprise as I was seeing the Loan Arranger.

I opened a savings account with £50. But my good intentions came to nothing, as I had soon withdrawn all the money again in small amounts as I needed them.

I kept a list of my withdrawals and balances, and on checking them, found that there appeared to be a pound missing. Can you explain?

	Withdrawal	Balance
I had £50. I took £25 out and left £25.	£25	£25
I then took £10 out and left £15.	£10	£15
I then took £8 out and left £7.	£8	£7
Next I took £5 out and left £2.	£5	£2
Finally I took £2 out and left nothing.	£2	£0

So on the left we have £25 + £10 + £8 + £5 + £2 = £50.
But on the right we have £25 + £15 + £7 + £2 = £49.

What happened to the extra pound?

Answer on page 150

12. In proportion

Two numbers, 100 and 20, are in the ration of 5 to 1.
What same number must you add to each, so their ratio
is 3 to 1?

Answer on page 150

13. On the square

100 is the square of 10, as $10 \times 10 = 100$. Can you find
2 numbers whose squares add to 100? If it will help you,
the numbers are in the ratio of 3 to 4.

Answer on page 150

14. Spare me more squares

Can you find 2 numbers whose squares add to 1000?
The numbers are in the ratio of 1 to 3. To help, we've had
one of the numbers already.

Answer on page 150

15. Shanks's pony

A horse travels half its journey at 12 mph. Then his
master loads him up and he travels the other half of the
journey at 4 mph. What is the horse's average speed?

Answer on page 150

16. Has the penny dropped?

You have three 10p and three 5p pieces. You put them
into three boxes, two coins in each. The boxes are labelled
10p, 15p and 20p and each box contains 10p, 15p or 20p.

But not a single box label is correct!

Each box has a slot, so by shaking it you can get a coin to come out. Can you determine what each box contains by shaking just one box to see one single coin?

Answer on page 151

17. Dozen really matter

In the days before we all went metric people often counted in dozens instead of in tens. So, what is the difference between six dozen dozen and half a dozen dozen?

Answer on page 151

18. It's people that count

This is a magic trick you can perform with two people and a large pile of counters. You turn your back and ask them to follow these instructions:

First ask person A to take a number of counters greater than 5.

Secondly, ask person B to take 3 times as many as A.

Now ask person A to give B 5 of his counters.

Now ask B to count A's counters and give A three times that number.

Now you amaze them both by declaring exactly how many counters B has left.

What will that number be, and how does it all work?

Answer on page 151

19. The missing numbers

Around a hundred years ago, missing number puzzles were very popular. Here is quite a simple one.

$$
\begin{array}{r}
9** \\
5*\,\overline{)*****} \\
**3 \\
\hline
*** \\
**5 \\
\hline
*** \\
**1 \\
\hline
0
\end{array}
$$

Can you fill in the missing numbers?

Answer on page 152

20. Never ask a lady her age

The *Women's Almanac* was a magazine that ran from 1704 until 1841. It often published puzzles and this one came in the form of a short poem.

> *If to my age there added be,*
> *One half, one third and three times three,*
> *Six score and ten, the sum you'll see,*
> *So tell me please, what age I be?*

Can you find the answer?

Answer on page 153

21. A calculated guess

Take a calculator and key in a three-digit number.
Now key in the same number again to make a six-digit
number, e.g. 123,123.

Let's have a little gamble. What are the chances that this
number is divisible by 7? The odds are 6 to 1 against.
Let's try it: 123,123 ÷ 7 = 17,589 with no remainder. That
was lucky.

Now let's have a bigger gamble. What are the chances
that the new number (17,589) is divisible by 11? The
odds are 10 to 1 against. Let's try it: 17,589 ÷ 11 = 1,599
with no remainder. That was even luckier.

Now let's really go out on a limb. What are the chances
that this new number is divisible by 13? The odds are 12
to 1. Let's try: 1,599 ÷ 13 = 123 with no remainder. Wow!
How lucky is that?

But now notice that the number we have ended up with
is 123. This is the number first chosen and entered twice
at the start. Can you explain what has happened here?

Answer on page 153

22. Triple tapping

On a calculator, tap in any two-digit number and repeat
it three times. If, for example, you chose 28, you would
now have 282,828.

Now see if that number will divide by 3 and leave no remainder.

Now see if the new number will divide by 7 and still leave no remainder.

Now see if this new number will divide by 13 and still leave no remainder.

Now see if this new number will divide by 37 and still leave no remainder.

Now check the number you are left with. Can you explain what has happened and why?

Answer on page 154

23. Gifted sons

It is Christmas and two fathers each decide to give their sons money as a present. One dad gives his son £100. The other dad gives his son £50. When they count up, the two sons find they only have £100 between them. Explain.

Answer on page 154

24. Think of a number

The number 12 is 4 times the sum of its digits (3). Can you find:

A whole number equal to exactly twice its digits?

A whole number equal to exactly three times its digits?

Two other numbers equal to four times their digits, as well as 12 and 24?

Answer on page 154

25. Chinese crackers

This is a very ancient Chinese problem, perhaps as much as 3000 years old.

No way? How? Could he have been the Chinese chap who thought it up? Perhaps not? Here's the puzzle …

There is a mystic number.
Divided by 3 will leave a remainder of 2.
Divided by 5 will leave a remainder of 3.
Divided by 7 will leave a remainder of 2.

What is the number?

Answer on page 154

26. Three little maids – in a flat

Remember the film *Boeing Boeing*? The story was about a chap who had three air hostess girlfriends, all at the same time – the cad! A lucky cad, but still a cad.

If any of them found out, they would probably kill him, so the trick was to make sure they never met. This wheeze worked as long as only one was in the city where he lived, at any one time.

In our puzzle there are no boyfriends involved. Three air hostesses share a flat. They all come home at different intervals:

One comes home every 5 days.
One comes home every 4 days.
One comes home every 3 days.

They are all at home today and leaving again. In how many days will they all be at home together again?

Answer on page 155

4

Easy Peasy Puzzles and Catchy Watchy Questions

Many of the following are old puzzles that occur again and again. But it would be a pity to omit them, as many readers may not have met them before.

Most are also really easy peasy, and some are more trick questions than puzzles. But if you are meeting them for the first time, or have forgotten the first time around experience, you might find them amusing.

They only require a few seconds thought – or do they? Be careful, they can easily trip you up.

1. Off the record

The lovely Norman, a.k.a Fatboy Slim and husband of my daughter Zoe, still takes the majority of his gig music straight from vinyl discs. I bet there are some youngsters out there who don't even remember vinyl records. But for those who do, here's a question.

How many grooves were there on a normal 12-inch vinyl LP?

Answer on page 156

2. Look at it another way

Visiting Zoe in Brighton, we often walk along the prom
and back. Wherever we go, we have to come back the
same way – I envy people who can do round trips.
Here's a roundabout sort of puzzle.

A man jogs each weekend to keep fit. He has sorted out
a circular sort of route, which takes him 80 minutes at
his same steady jogging speed. However, one week,
just for a change, he ran around the other way. When he
checked his watch at the finish it had take him – shock
horror – 1 hour and 20 minutes. Why?

Answer on page 156

3. Stock is money

You could get these in a shop that sells fork handles or
even four candles. One costs 50p, 14 costs a £1, but you
can get 957 for £1.50. What are they?

Answer on page 156

4. Crows' feat

There are 21 crows in a field. The farmer shoots 3. How
many crows are left in the field?

Answer on page 156

5. You have the gift

You have 8 apples and give all but 3 away. How many
have you left?

Answer on page 156

6. Uncorking a corker

You want to buy a bottle and a cork. They cost 20p. The bottle costs 19p more than the cork. How much does each one cost?

Answer on page 156

7. Eggsasperation

This is a puzzle my Grandmother tried on me: should you say, 'The yolk of an egg is white' or 'The yolk of an egg are white'?

You should say neither. The yolk of an egg is yellow. Doh.

Here's another puzzle. This yokel had an egg for breakfast every single day. But he didn't buy any eggs. He didn't steal any eggs. He didn't borrow any eggs. And he didn't keep chickens.

So where did he get his eggs?

Answer on page 157

8. Cutting it fine

Why would a barber who supports Manchester United rather cut two Liverpool supporters' hair than one Manchester City supporter's hair?

Answer on page 157

9. A catfish conundrum

If a cat and a half eat a fish and a half in a day and a half, how many fish would 7 cats eat in a week and a half?

Answer on page 157

10. Ark at this

How many animals of each sex did Moses take onto the Ark?

Answer on page 157

11. Tweedling their sums

Tweedledum and Tweedledee were twins. They looked alike, spoke alike, acted alike – they were almost totally alike. But one was heavier than the other.

Tweedledum said to Tweedledee, 'The sum of your weight and twice mine is 151 kilograms.'

Tweedledee replied to Tweeedledum, 'The sum of your weight and twice mine is 152 kilograms.'

How much did each one weigh?

Answer on page 157

12. A word to the wise

Which word is almost always pronounced wrong by Radio DJs?

Answer on page 157

13. Under a cloud

If it is raining at midnight, what are the chances that the sun will be shining in 72 hours?

Answer on page 158

14. Private lives

Two soldiers were sitting in the barracks, a colonel and a private. The private was the son of the colonel. But the colonel was not the father of the private. How is this possible?

Answer on page 158

15. Have a break, have a …

Chocolate bars come in all shapes and sizes. You can have one made up of 2 segments or even 48 segments. Let's assume you have a chocolate bar made up of 20 segments. What is the minimum number of snaps required to break the bar into 20 individual pieces?

Answer on page 158

16. Mind-numbing numbers

Can you think of two whole numbers which, when multiplied together, make 7?

Answer on page 158

17. Only the half of it

Divide 50 by ½ and add 8. What is the answer?

Answer on page 158

18. Percent to get you!

One morning, the boss announced to his staff, 'Due to a temporary decline in business, I want you all to take a 20% wage cut.' There were strong murmurs of discontent around the office.

'Don't worry,' said the boss, 'as soon as things pick up, I will increase all your wages by 20%.'

So were they happy and should they have been happy?

Answer on page 159

19. To coin a phrase

I have two coins in my pocket that add up to 15p, but one is not a 10p. What denomination are the two coins?

Answer on page 159

20. Got you surrounded

How can you draw a circle around someone, so they cannot jump out of it?

Answer on page 159

21. The magic money clip

Rather than a wallet, some people use a money clip to hold bank notes together. A simple paper clip would do as a money clip. But can you use a bank note, to magically link two paper clips?

Answer on page 159

22. In the cold light of day

A polar explorer is marooned all alone in his tent. His friends have gone for help, but they have now been gone several days.

More days pass as he waits in vain for the anticipated rap on his flap. But no rap comes. Provisions are running low. He is now desperately desperate, and getting colder and colder. He has a gas stove, a paraffin heater, a candle, a wax taper and only one match. In desperation he tries to think … which should he light first?

Answer on page 160

5

Geometric Shape and Angle Puzzles

Geometry is my favourite branch of mathematics. I have always felt that in reducing the amount of geometry taught in British schools, we have weakened both the number and quality of mathematicians that the system is producing.

Throughout the nineteenth and early twentieth centuries many great figures claimed that their powers of thinking and intellect would never have been so well-developed had not Greek geometry shown the way. Geometry is rather like a detective whodunit mystery. You are given clues like the size of an angle or the length of a line, then with a little more information suddenly you can put all the evidence together and come up with a solution that at first seemed to be beyond your power and ability.

So, I felt I had to have a chapter in this book which just played with shapes – a chapter on geometry.

The great French mathematician and philosopher, Rene Descartes (say 'day cart') wrote a book called *The Geometry*. It showed and explained that engineers could measure anything using geometry and a few simple rules. How powerful is that? Well, it was a major mathematical influence in causing the Industrial Revolution a hundred years later.

The introduction to *The Geometry* gave two examples of the power of geometry. I have featured them as questions 22 and 23.

So, here are a few puzzles for the budding geometers amongst you.

1. Halving the pain

This puzzle was a favourite of Lewis Carroll.

I have a square window, but it is too big. How can I make a new square window to fit inside the big square window, one which is exactly half the size of the big square window?

Answer on page 160

2. A measure of your ingenuity

Imagine that you are standing on the banks of a straight river or canal. Across from you is a signpost on the opposite bank. To your right, along the bank, is a tree. You have nothing to measure with. How can you find a pretty accurate distance to the signpost?

Answer on page 161

3. Try angling

In a field there is a triangular lake, where you could try angling – ha ha. It's a perfect triangle with three straight edges and sharp corners. Walking around it, you complete a circle and so have gone through 360 degrees. But you have gone around a triangle which has angles that add to 180 degrees? How do you explain that?

Answer on page 162

4. Square dogs

We have a perfectly square field, 100 metres on each side. We have four hunting dogs, one at each corner. These dogs will hunt anything. If there are no other animals about, they will hunt each other.

Suddenly they set off and race towards another, all at a constant and identical speed. Each dog heads for the one that was to his right. They keep going full pelt, until they meet. They all run at exactly 4 metres per second.

So, how far will they run, where will they meet and after how long?

Answer on page 162

5. Cutting corners

A normal chess board has 2 opposite corner squares removed. Now, instead of 64 squares, it has 62 squares. You have 31 dominoes, each able to cover 2 squares on the chess board. Is it possible to cover all the squares on the board using the 31 dominoes?

Answer on page 163

6. House hunting

Here's a thought – do people who go house hunting, need a very big net?

Let's assume you are approaching a perfectly square house. You stop some distance away. What are the chances that you will be able to see two sides of the house?

Once you have thought about that one, you might like to consider the odds of how many sides you will be able to see if you are approaching a three-sided house, or a five-sided house.

Answer on page 164

7. Along the same lines

You have two lines, each 1 unit long. They meet at one end to form the two equal sides of an isosceles triangle. What length must the third side be, if the triangle is to have the largest possible area?

Answer on page 166

8. Strictly for cubes

Here is a cube made of 3 layers of 9 smaller cubes, like a Rubik's cube. There are 27 cubes in all. Note that alternate cubes are coloured black and white.

Say that you made 13 double cubes, each formed from a black and a white cube. You could then assemble the double cubes into a 3 × 3 x 3 cube, as in our picture, leaving 1 cube empty.

Could you arrange them so that the empty cube was the very centre cube?

Answer on page 166

9. Strictly for squares

Here are two unequal squares, side by side. There is a simple way to draw two lines and divide the squares into five pieces that will reassemble into one large square. Can you find the solution? As a hint, the two lines need to meet at the right angle.

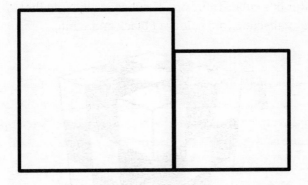

Answer on page 167

10. Just passing through

Here is a diagram of 12 squares, with a diagonal line running from opposite corners.

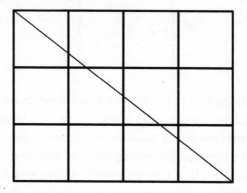

How many squares does that line pass through? No, the answer is not 6.

Answer on page 168

11. Getting the drop on the count

Compte de Buffon, an 18th century French mathematician, did an experiment. He dropped lots of needles on a planked floor and counted which touched or overlapped a line and which did not. Doing this, he discovered a fact about circular maths.

So try it. Get a box of pins – or matches or needles, all the same length. Now rule parallel lines on a large piece of paper. Each line must be exactly twice the length of the pins apart. Now start dropping the pins onto the paper at random.

BALL OF CONFUSION

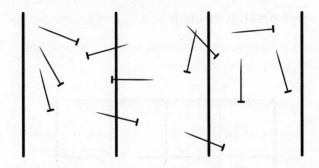

Record the number you drop and count up those that land touching (or crossing) a line. Divide the smaller number into the larger number and note the result. Try it several times, still recording the total dropped and the total touching a line. Can you spot how the result has a connection with circular maths? Can you give that figure a name?

Answer on page 168

12. Look out! There's a bear behind!

One day a man went out to shoot bears. He wasn't a hunter, he was a photographer and with his camera, he set off just after breakfast to find bears.

He walked 5 miles directly south, but didn't see a single bear, or a married one come to that. So he turned 90 degrees left and walked 5 miles east, where he saw a bear. But the bear saw him and began chasing him. So he turned 90 degrees left and ran directly north, with a bear behind – him. The bear gave up the chase, but the man kept running until he had covered 5 miles. Now, to his surprise, he was back at his camp.

Can you explain how this could be and tell me, what colour was the bear?

Answer on page 169

13. Same question from the other end

Following on from the previous question, could our photographer have made exactly the same journey, including exactly the same distances and turning points, and have seen a penguin?

The answer is yes – but can you explain how, and exactly where his journey might start from?

Answer on page 170

14. By the light of the silvery lune

Hippocrates of Chios was the chap who thought this one up. In doing so, he was the first person to accurately measure the area inside a curved shape – but he couldn't have done it without Pythagoras.

Pythagoras said that, 'In any right-angled triangle, the square on the hypotenuse always equals the sum of the squares on the other two sides.'

But that doesn't just go for squares. Any shape drawn on the hypotenuse of a right-angled triangle will have the same area as the sum of the two identical to-scale shapes on the other two sides.

So, you could have an elephant on the hypotenuse and smaller to-scale elephants on each of the other sides. The area of the large one would equal the combined areas of the other two.

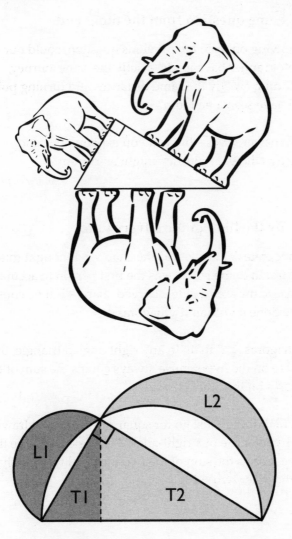

In the diagram, semicircles have been drawn on the hypotenuse and on the other sides. This has produced two shapes, curved on both sides like new moons, which is why they are called lunes. The question is, can you tell me the area of each of the lunes?

Answer on page 171

15. An elliptical orbit

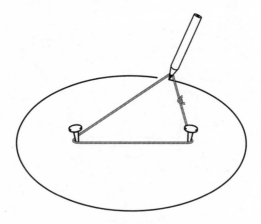

The simple way to draw an ellipse is to stick two drawing pins a certain distance apart and make a loop of string that, when stretched out, is wider than the two pins. You then place the loop around the pins and pull it taught with a pencil. Now sweep the pencil round, keeping the string taught, to draw a perfect ellipse.

But how do you draw an ellipse of set dimensions? That is the question. Can you find a way to draw an ellipse exactly 12 cm wide and 8 cm tall?

Answer on page 172

16. Upsetting Lewis Carroll's applecart

Lewis Carroll once imagined a four-wheeled cart with
oval or elliptical wheels. What would happen as it was
pulled along? The cart would roll and pitch like a boat
on a wavy sea. Then Carroll asked this question – how
would you set four oval wheels so that the cart made the
smoothest ride possible?

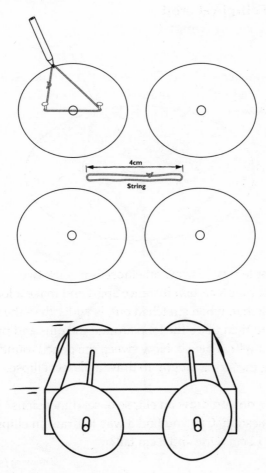

You could easily make a cart with four elliptical or oval wheels – simply copy the diagram. You could use a small box for the cart and ballpoint pens for axles. Or you could think about the problem and just use your imagination. But what wheel setting do you think would make the smoothest ride?

Answer on page 173

17. Curves of constant width

We are all used to seeing round wheels, and it's difficult to imagine wheels any other shape that would give a vehicle a smooth ride. But in fact there are an infinite number of shapes that would do this. All you need is for your shape to always have a constant width.

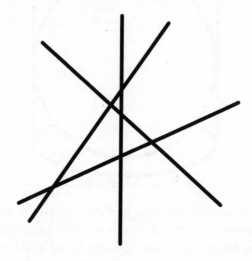

So, here are four lines, each of which crosses the other three. Can you, using a pair of compasses, make a shape where its width is always the same? Draw your own four lines and try it.

Answer on page 174

18. Of grave concern

Before Archimedes died, he asked that a sphere within a cylinder be engraved on his tomb. Centuries later, when searching for his tomb, another mathematician found the engraving and knew he had found the tomb of the great man. But why had Archimedes chosen this design?

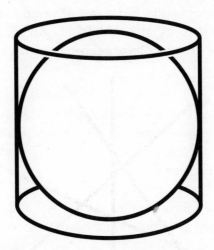

It was because Archimedes had discovered the answer to this question. In the diagram, what is the relationship between the surface of the sphere and the surface of the sides of the cylinder?

Answer on page 175

19. A new train of thought

UK high speed trains travel at 125 mph. Continental expresses go even faster. Can it be possible that at high speeds, certain parts of a train are going backwards?

Answer on page 176

20. The quickest way down

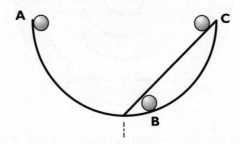

This drawing is of a cycloid turned upside down. At the centre bottom of the curve is a point we will call X. A ball could run down the curve of the cycloid or down the straight line to X.

I have chosen three possible starting points. Ball A will run from the top of the cycloid down the curve to X. Ball B will start a good distance down the slope and run along the curve to X. Lastly, ball C will run down the straight line, directly towards X.

But which ball will arrive at X in the shortest time?

Answer on page 177

21. Target practice

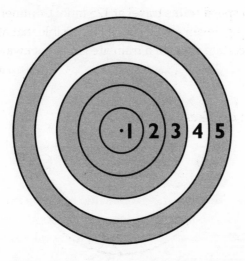

This target has five concentric rings, all 1 unit in width. The inner three rings and the outer ring have been shaded. Which of these two shaded areas do you think has the largest area? Remember, the area of a circle is πr^2 – that is all you need to work out the answer.

Answer on page 178

22. The Mesolabe Compass

This idea features as an introduction to Descartes' *The Geometry*. It was originated, as far as I know, by Eratosthenes in Alexandria around 200 BC. He called it the Mesolabe Compass. It asks the question, can you use two rulers to multiply or divide any two numbers?

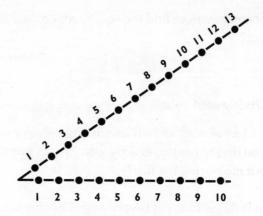

Draw two lines, one horizontal and one rising from one end to form an angle. The actual angle doesn't matter, but around 40 degrees works fine. Mark off both lines in centimetres. With this device, you can multiply any two numbers and divide any two numbers.

Can you explain how?

Answer on page 179

23. Getting to the root of the problem

Square roots were always a major teaching problem when I was at school. However, Descartes found in Euclid's geometry the simplest possible method. He used it as the second simple yet powerful example in his introduction to *The Geometry*.

You have a base line marked off with centimetre units and numbered from 0–20. Can you use this line and a

pair of compasses to find the square root of, say, 9 for
instance?

Answer on page 180

24. Divide and rule

Draw a line of any random length on a piece of paper.
Can you divide the line exactly into 7 equal parts,
without measuring the line?

With a little geometry it is very easy. The clue is in the
Mesolabe Compass puzzle, question 22 in this chapter.

Answer on page 182

25. The spider and the fly

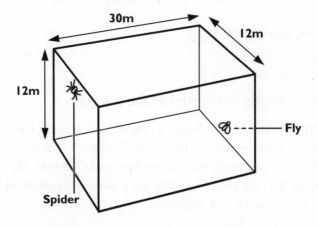

In an old mansion was a perfectly rectangular room
30 metres long. The end walls were 12 metres high and
12 metres wide. The room was empty save for a solitary
fly and a hungry spider.

The fly stood motionless halfway across the end wall and 1 metre above the floor. The spider was diametrically opposite, halfway across the other end wall. He was exactly 1 metre from the ceiling. Being hungry, he wanted to get to the fly in the quickest time possible, so he wanted to take the shortest route along the walls that he could.

Can you find the shortest possible route from spider to fly? You can draw the room opened out like a cardboard box to help you.

Answer on page 183

6

Dear Old Faves from Days Gone By

Many puzzles have occurred time and time again in puzzle books down the ages. I decided quite early on in writing this book that I would not include those, but try to stick to puzzles that were fresher and not too widely known. However, so many of those old puzzles are very dear favourites and each time I edited one out, it hurt.

I have been entertaining for around 50 years, and I once got a mention in a daily newspaper which said, 'Old fave, Johnny Ball was there.' I can tell you that I certainly wasn't upset by that.

So, here is a collection of some very dear 'old favourites'.

For some readers this chapter will be a walk down memory lane, but there is always a new group of young puzzle enthusiasts in the pipeline and for them to not experience the 'old faves' would, I think, be very sad indeed. Enjoy.

1. The sweet truth

You have three jars of sweets, which you cannot see into.

One is labelled 'jelly babies', one labelled 'wine gums' and one labelled 'mixed'. But all the labels are wrong.

Can you find out exactly what is in each jar, by taking just one sweet from one jar? If so, then which jar?

Answer on page 184

2. A later edition

Earlier in the book I asked a question about missing pages in a newspaper. This puzzle is slightly different. There are a total of 28 pages in a newspaper. If page 19 is missing, what other three pages are also missing?

Answer on page 184

3. Half full, half empty

There are six glasses in a row. The first three are full of wine, the last three are empty. By moving only one glass, can you arrange a row of full and empty glasses lined up alternately?

Answer on page 184

4. Beauty is only skin deep

Two beauticians run a skin and body care shop. One day they have three customers, each needing a face pack and a manicure in a hurry. A face pack takes 15 minutes, and a manicure takes 5 minutes. What's the quickest time in which they can get the job done?

Answer on page 184

5. A leap of faith

A frog falls down a 10-metre well. It is not too worried as it can jump 3 metres high. But the walls of the well are slippery and after each jump, it slowly slips back 2 metres. It is now exhausted. It finds a spot to cling to and sleeps for the rest of the day. The next day it takes another jump up 3 metres, but slides back 2 once more.

How many days does the frog take to jump out of the well? Well?

Answer on page 185

6. Don't get this one rung

A boat is moored to a jetty on the river Thames with a rope ladder over the side. There are 12 rungs of the ladder showing above the water, each one 30 cm apart. The tide is starting to come in and rises 30 cm every hour. How many rungs of the ladder will be visible after 4 hours?

Answer on page 185

7. Combined assets

This funny conundrum was set by Lewis Carroll, and he wasn't born yesterday.

Alf and Bill each had £500 which they took and deposited in the bank. Next day they went back to the bank and had well over £1,000,000 between them. How could that be?

Answer on page 185

8. The last word in puzzles

Take a letter, or rather, take four letters all consecutive in the alphabet. Which four consecutive letters of the alphabet, when rearranged, form a word, or two? What word or words?

Answer on page 185

9. Is there a short cut?

How do you catch mice? Put cheese on your tongue and wait with bated breath.

Imagine a cube of cheese, like a Rubik's cube. Imagine it is made of 27 small cubes. How many cuts are needed to release the centre cube?

Answer on page 185

10. A rather nasty turn

On the road to eternity, I am told (I haven't been myself), there is a fork in the road. One branch leads to Heaven and the other to Hell.

There are also two sentries who guide the way. One is an angel and the other is one of the devil's disciples. However, they both look identical and their uniforms are at the cleaners, so you don't know which is which. However, angels always tell the truth and the other lot always lie. By asking just one question, can you determine which is the road to Heaven?

Answer on page 186

11. An up and down kind of existence

A monk lives on a river bank in a valley. Once a week he must travel to the monastery at the top of the Lonely Mountain. He sets off at dawn, as the road is steep and long, and he arrives as the Sun sets. Next morning he has a lie in and leaves for home at 10 am. As the road is downhill all the way, he arrives home at around 3 pm.

What are the chances of the monk being at exactly the same point on the road, at exactly the same time on both days?

Answer on page 186

12. Bending the rulers

Find a school ruler and make a quick, simple paper copy of it – just mark off the centimetres. Now fold the paper ruler any way you like while laying it along the school ruler.

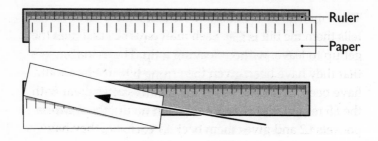

The only rule is that the paper ruler must all be on top of the school ruler. Can you fold the paper ruler on the school ruler, without two identical points along both rulers still being in line vertically?

Answer on page 186

13. Losing one's bearings

You have two identical maps. One is laid out flat on a table. The second is crumpled up into a ball and placed on the flat map. What are the chances that some particular spot, will be in line vertically on both maps? But also, what are the chances of some identical spot being in line vertically if the two maps are of a different scale?

Answer on page 187

14. Short changed

This puzzle really has stood the test of time. I remember it as a young lad and I still hear it trotted out at least once every year.

Three men go to a restaurant. After the meal the waitress tells them the bill is £30. Each man pays £10 and then they get up to leave, without leaving a tip. The manager sees that they have been given the wrong bill. Theirs should have been £25. So he sends the waitress after them with the £5 refund. But because they left no tip, the waitress pockets £2 and gives them back £1 each. So, they have now paid 3 × £9 = £27 and she has £2 in her pocket.

What happened to the extra £1?

Answer on page 187

15. Talking telephone numbers

I did this trick in one of my early BBC *Think of a Number* programmes in around 1980. I appeared to memorise a telephone directory, but I actually only memorised one number. It was the ninth number down on page 108. I then performed the trick. Try it yourself now, and then try it on a friend later.

Choose any three-digit number with the end digits different by at least 2. Now reverse the chosen number and take the smaller from the larger of the two. The answer is another three-digit number. Reverse that and add the two numbers. The answer had four digits.

Now, find the page in the directory corresponding to the first three digits. Count down that page according to the last digit. You will now be on page 108, and looking at the ninth number down. But how come?

Answer on page 187

16. The leaping loop

This is a trick you can perform with an elastic band, anywhere, any time.

Loop an elastic band around the index and middle fingers of your hand.

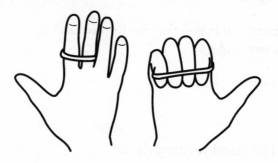

Then make your hand into a fist and straighten the fingers. The band is now magically looped around the other two fingers.

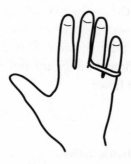

Your audience will suggest the band simply jumps over the ends of your fingers. So, now you repeat the trick but this time make it look bafflingly impossible.

Place the band around your index and middle fingers again.

Then wrap a second band around your four finger ends.

Now make your hand into a fist and then straighten out the fingers.

The band still makes the now impossible leap to the other two fingers.

But how?

Answer on page 188

17. The mind reading rip-off

Announce that you can mind read and are able to know what someone writes down in secret. You need any number of people up to nine. You then tear a piece of paper into nine pieces. Everyone writes something in secret and hands the pieces to someone who scatters them in front of you. You now say that you cannot read everyone's mind, but you can read one person's mind. You then choose one of the group and identify what that person wrote. But how?

Answer on page 189

18. All minds think alike

Announce that when one person thinks of something, those also in the room will tend to think the same thing. To prove this, you will conduct an experiment. Ask those assembled to:

Think of a number, double it, add 8, divide by 2 and take away the original number.

Now ask them to convert the number into a letter of the alphabet, 1 = A, 2 = B etc.

Now ask them to think of a country beginning with their chosen letter.

Now, ask them to think of an animal beginning with the second letter of the country.

Lastly tell them to think of a colour that you would associate with the animal.

Then announce that other than in zoos, there are no grey elephants in Denmark. Now, how does that work?

Answer on page 189

19. Where on Earth?

Where in the world could you see the sunrise twice in any 24-hour period?

Answer on page 190

20. This might give you a furrowed brow

This is a very ancient puzzle.

A Babylonian ox is ploughing a field. The field is 200 cubits by 200 cubits. The ox and plough are 4 cubits wide. How many tracks will the ox leave when it has ploughed the field?

Answer on page 190

7

Even More Thoughtful Thinking Puzzles

The writer of *Madame Bovary*, Gustave Flaubert, when a young man, sent this puzzle to his sister who had just started studying mathematics at university.

> *Since you are now studying geometry and trigonometry, I give you this problem: A ship sails the ocean. It left Boston with a cargo of wool. It grosses 200 tons. It is bound for Le Havre. The mainmast is broken. The cabin boy is on deck, there are 12 passengers aboard, the wind is blowing east-north-east. The clock points to 3.15 in the afternoon. It is the month of May. How old is the captain?*

Don't give it too much thought – it is nonsense and meant as a joke, implying that though his sister seemed to be enjoying it, he hated maths, puzzles and puzzling.

But for true puzzlers, the more complex the puzzle, the more worthwhile and rewarding the time spent in puzzling out the answer. So, don't give up easily on the puzzles in this chapter – persevere. Only if your brain goes completely numb should you look up the answer. Have fun!

1. Jobs for the boys

Alf, Bill and Charlie each have two jobs, in these austere times. Their jobs are driver, shelf stacker, musician, painter, gardener and hairdresser. From the following facts, can you sort out which chap has which two jobs?

1. The driver offered the musician a lift.
2. The musician and the gardener went fishing with Alf.
3. The painter bumped into the shelf stacker in the supermarket.
4. The driver went out with the painter's sister.
5. Bill owes the gardener £5.
6. Charlie beat Bill and the painter at poker.

Answer on page 190

2. If truth be told

I once had a holiday on the Isle of Confusion. Oh yes I did! On the island live three tribes. One tribe always tell the truth. Another tribe always lie. The third tribe, known as the Eachwaybets, alternate between lying and telling the truth.

As luck would have it, I met three of them, one from each tribe. Speaking with my mouth, I said, 'Which tribe do you each belong to?'

Chap A said, 'C always tells the truth.'

Chap B said, 'A always tells the truth.'

Chap C wouldn't say a blooming thing, but that didn't matter, as I now already had all the information I needed to work out who belonged to which tribe. Can you sort out who was from which tribe?

Answer on page 191

3. Angels with dirty faces

Two school boys are playing around on a shed roof. It collapses and they fall through the roof and land on the muddy floor. They both get up and are okay. However, one now has a muddy face and the other a clean face.

They are now late for lessons and race back to class. The one with the clean face, however, nips into the toilets to wash. Explain.

Answer on page 191

4. This could drive you dotty

This is a puzzle you need to think about, rather than physically do it.

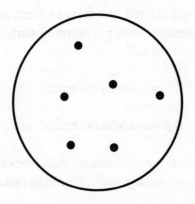

Imagine you have a circle in which you place an even number of dots. Lots of dots. How many? Our picture has just 6. But let's imagine there were 500. Would it always be possible to draw a line through the circle that would divide all the dots exactly into two halves? If so, how?

Answer on page 192

5. A king's kith and kin

King Solomon had a thousand wives. He serenaded them daily,
What's the use of a thousand wives, if you've only got one ukulele?

A king was once very worried about the many children he would bring forth from his many wives and what sex they might be. He needed boys first and foremost, to inherit and run his kingdom. But he also needed daughters to marry princes from other kingdoms, to help form powerful alliances.

So, to try to ensure the right mix of offspring, he declared, 'When any of my wives gives birth to a boy, she will stop having children. When a wife produces a girl, she can have more children, until she produces a boy. This will ensure I have many boys to inherit my kingdom but also some girls to marry wealthy princes.'

Was the king's thinking correct, and what would be the overall ratio of boys to girls in his children?

Answer on page 193

6. Happy birthday to you two

This is a puzzling question that crops up from time to time. So, it may not be new to you, but I need to set it down here, as a lead up to the next puzzle.

How many people need to be gathered in one room for the chances of 2 of them having the same birthday, to be better than an even chance? If it is new to you, give it some thought and hazard a guess?

Answer on page 193

7. A hair-raising question

How do you avoid falling hair? At the very last minute, step out of the way.

People have a varying number of hairs on their heads, from over 100,000 to 0. Amazingly, genuine blondes tend to have most hairs at around 140,000. Most people tend to lose hairs as they get older. I still have the hair I had 40 years ago. It's in a shoe box under the bed.

Let's assume that the average person has 80,000 hairs on their head. How many people do you need in one place before the chances are better than even that two of them will have the same number of hairs on their heads? From the previous question, see if you can guess at the answer.

Answer on page 194

8. The answer is buried in the sand

I devised this puzzle thanks to a lady called Liz Weeks, who sent me a similar puzzle via Zoe.

When Zoe, Nick and Dan were kids, Di and I had so many wonderful holidays with them. In those days I was on TV most weeks with shows like *Think of a Number*, which meant I was quickly recognised by other families. I didn't mind at all.

However, we preferred to avoid the crowds. On various Mediterranean islands, we would hire a car, meander down roads that were little more than dirt tracks and find deserted beaches. We would not see another human soul in an entire day – it was bliss.

One of the games all of our kids loved was burying me in the sand, leaving my head sticking out of course – at least most times. It was all good fun and they usually came back and dug me up again before it was time to leave.

Zoe, being the oldest, working on her own, could bury me in 10 minutes. Nick, who was next eldest, took 15 minutes burying time. Dan was the youngest and could do it, but it took him 30 minutes.

One day, they decided to all work together to bury me. Assuming they all worked at their usual rate, how long did it take them to bury me?

Answer on page 195

9. He hadn't got a sausage?

A businessman found himself sharing a railway carriage with two young men and an old chap in the corner, who was thankfully asleep. One of the young men had with him 5 sausage rolls. The other had 3 sausage rolls. The train broke down, as they do from time to time and the men were suddenly stranded.

The old man was still asleep, so the other three shared the sausage rolls by dividing them equally between the business man and the two young fellas. When they finally arrived at their destination, the businessman checked his pockets. He found he had £8 in change, which he gave to the young men as he left. That was when the row started, which also woke up the old man.

The 5 sausage roll lad insisted he should have £5 and the other chap £3. The 3 sausage roll lad reckoned they should both have £4. After more argument, the 3 sausage roll man suggested he would accept £3. That was when the wise old man sitting in the corner butted in.

He explained to them the actual fair division, based on their contributions. Now, what was that fair division?

Answer on page 195

10. Weigh above your head?

When his boss came back from lunch, the goldsmith's assistant said, 'A woman just came in and delivered 12 gold coins. But she said one is a dud.'

'Counterfeit?' asked the goldsmith.

'I think she had two.' said the stupid boy,

'Did she say whether the counterfeit coin was heavier or lighter?' asked the boss, wearily.

'Uh, I don't remember.' said the stupid boy.

So there, in front of the goldsmith, were 12 gold coins and his balance scales. He had no weights, but he didn't need weights. He identified which was the dud coin and whether it was heavier or lighter, and he did it in just three weighs.

But how did he do it?

Answer on page 195

11. A monkey puzzle

In the Olde Flour Mill in the village of Yerkiddinme there is a rope which passes over a pulley wheel, which is fastened to the ceiling. On one end of the dangling rope is tied a sack of flour. On the other dangling rope, at the same level as the sack of flour, is a monkey.

The monkey weighs exactly the same as the sack of flour, so they are balanced. We can assume the rope does not suffer from any friction passing over the pulley wheel. The monkey starts to climb the rope. What happens to the sack of flour and to the monkey, as the monkey climbs the rope?

Answer on page 197

12. Blind faith

Three people are blindfolded. On each one's forehead a white card is stuck. They are told, 'The cards are either black or white but not all the cards are black.' They are also told that they should be able to identify the colour of their card. As soon as they know the colour of their own card, they must say what it is.

The blindfolds are removed and after some thought all three say their card is what colour, and why?

Answer on page 197

13. Hey diddle diddle, there's a hole in the middle

A machine operator has a ball. He drills a hole directly through the centre of this ball. When he has finished, the height of the wall of the hole is 6 cm. What is now the volume of this ring with the hole in the middle? It doesn't seem that I have given enough information for you to find the answer. But I have.

Answer on page 198

14. An amazing chain of events

The Lady Dowager is travelling abroad on her own when a dastardly thief steals her wherewithal. She finds an inn and sends a message for someone to come to rescue her. Meanwhile she has to pay the inn keeper.

Luckily, hidden in her clothing, she has a length of gold chain with just 7 large links.

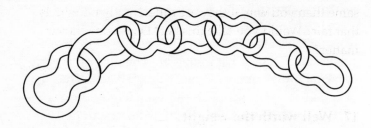

The inn keeper agrees that he will accept 1 gold link per day as payment. Someone will surely arrive to take her home during the next week, but she does not know when. So, how many links does she need to cut, so she can pay a link each day?

Answer on page 198

15. Boys will be boys and vice versa

A boy and a girl were at a wedding reception, chatting together, as you do. The one with black hair said, 'I am a boy.'

The one with red hair said, 'I am a girl.'

Now, at least one of them is lying, so who is actually who?

Answer on page 199

16. How card sharp are you?

For this puzzle you can shuffle off for a deck of cards, or just give it some thought.

I have here 2 red and 2 black cards, face down. I am going to ask you to turn two up. If they are both the same then you win, if they are different, then I win. Is that fair? Would you take this bet? Does your choice matter?

Answer on page 199

17. Well worth the weight

Three lads together weigh 160 kg.

Bill weighs 48 kg.
The boy wearing sandals weighs exactly 8 kg less than the heaviest boy.
Charlie weighs more than the boy with trainers.
Alf weighs less than the boy in school shoes.

Can you sort out who weighs what and what shoes they are wearing?

Answer on page 200

18. Ill-gotten gains

I remember once when I was in theatrical digs we had a landlady we called Mrs Pirate. Why? Because her black beard repelled all boarders! Here are some puzzling pirates.

Cutthroat Jack has two mates, Swashbuckling Sid and Peg-Leg Pete. They decide to share out their booty. No, not the baby's footwear. They are going to share their ill-gotten gains which amount to 1000 pieces of eight. A Spanish silver coin was worth 8 reales in pirate days. Here's the puzzle …

Had the shares of Cutthroat Jack and Swashbuckling Sid been halved, then Peg-Leg Pete would have received three times what he actually got. Had the shares of Swashbuckling Sid and Peg-Leg Pete been halved, then Cutthroat Jack's share would have been 50% more than it actually was.

So what, my fine buckoes, were their actual shares? It is not as hard as it sounds.

Answer on page 200

19. Age concern

So I said to this fella, 'Tell me, how old is your son?'
He said, 'My son is as many weeks old as my grandson is days.'
So I said, 'How old is your grandson?'
He said, 'As many months as I am years.'
So I said, 'How old are you?'
He said, 'Well our combined age is 100.'
I said, 'Oh, great.'

So what were their ages?

Answer on page 201

20. Crazy golf

This fascinating puzzle was originated by Henry Ernest Dudeney who was Britain's best ever puzzle thinker-upper. He was born in 1857 and was at his puzzling prime between 1880 and 1920. This is typical of his puzzles. In his day they measured in yards and most golfers still do today.

There is a very strange nine-hole golf course, just outside Yeravinmeon, where the holes are 300, 250, 200, 325, 275, 350, 225, 375 and 400 yards long. A brilliant golfer arrives who has a very strange game. He only ever uses two clubs, which always send the ball absolutely straight and reach an exact distance in yards.

Sometimes a shot goes too far, passing directly over the hole. Then he uses the other club to come back exactly to the hole. One day he used two clubs which sent the ball exactly 125 and 75 yards. The next day he used clubs that sent the ball 200 yards and 25 yards.

With one of these pairs he went round in one stroke less than with the other pair. But on the third day he used another two clubs which sent the ball two different distances. This time he produced the best score possible.

What were the two club distances that achieved the lowest score of all?

Answer on page 201

21. A cruising conundrum

A ship leaves Southampton every day at noon for New York. One day you board this ship. The crossing takes 5 days exactly. Every day a ship also leaves New York at noon for Southampton and takes 5 days. How many other ships doing this journey, will you see on your trip?

Answer on page 202

22. An age-old problem

At my age, I tend to talk about age less and less as my own age gets more and more. But here is an age-old problem, to scramble your brain. Alf is twice as old as Bert was when Alf was as old as Bert is now. When Bert is as old as Alf is now, their combined age will be 63. How old are Alf and Bert now?

Answer on page 203

23. One more river to cross

Earlier in the book I featured a three tourists and three vampires crossing puzzle. As you can imagine, it will be slightly more difficult with four of each.

Four tourists need to cross a Transylvanian river. On the other bank are four vampires, wanting to cross the other way. There is a canoe which can hold three people, but only one of each group can actually row. I assume the others will look on in oar.

It goes without saying, but I'll say it anyway … at no time in the boat or on either shore, must vampires outnumber tourists, or it's neck-puncturing time.

Can they all cross the river safely? Of course they can, but can you work out how? The first thing to ascertain is, 'Where must the boat start?'

Answer on page 203

24. The three colour map theorem

Here is a map of England, Wales and Scotland, with all the counties outlined.

It has been known for centuries that only four colours are required, to colour any map so that no two adjacent areas are of the same colour. This is known as the four colour map theorem.

However, the UK is surrounded by sea, which we will assume is blue. Can you colour this map so that only *three* colours are used to colour the counties with a coastline?

You can of course use blue for counties with no coastline, but no two touching counties can have the same colour. It must be possible, otherwise the four colour map theorem is wrong? Try it.

Answer on page 204

25. Everyone to the boats

This popular puzzle was first asked and answered by a British mathematician called Thomas Kirkman in 1850.

15 schoolgirls go on a 7-day holiday. Their headmistress hires 5 boats daily, which each take 3 girls. Over 7 days, each girl could sail with 14 different girls. In this way no girl would need to sail with any other girl twice. So, can you draw up the schedule for the girls and boats for the week? Sounds simple, doesn't it?

Answer on page 204

26. Josephus is not one of us

I have to apologize for the fact that this ancient puzzle has always had racial overtones. The original story seems to have been that Josephus was a Jew in a cave with 40 other Jews who were bent on self-destruction rather than being captured by the invading Romans and suffering a fate too horrendous to contemplate.

Being a mathematician, Jo got them all to stand in a circle. He would choose a set number and they would count around the circle. When the agreed number was reached, that person would be eliminated. The counting and elimination would continue until just one person was left. Of course in the end that person was Josephus, who survived to tell the tale.

Versions of this puzzle have taken many forms, often involving refugees on an overladen ship which needed to shed passengers to prevent it from sinking.

Often the puzzle involves two parties, always with equal numbers, each of who would not mind eliminating the others.

There is a Japanese version where the two groups are the children of two wives of the dying emperor. The Black Queen suggests the counting and elimination system and all the White Queen's children disappear one by one. When there is only one left, the White Queen pleads that the counting now goes the other way. This time all the Black Queen's children are eliminated and the White Queen's last remaining child inherits the throne.

So, opposite is a version of the traditional puzzle with 30 souls set in a ring. I have labelled the two groups As and Bs. They have been arranged by a mathematician who is clearly an A, hoping to eliminate all the Bs.

A certain single-digit number is chosen. Starting from the A person at the top centre, the counting moves around clockwise. Each person that the count ends on will be eliminated. If you choose the right number, all the Bs will go before any of the As.

So, can you discover what the magic number is? Can you also say who would be the last B and the first A to go?

But now see if the Japanese version works with this layout. After 14 Bs have been eliminated, use the same number but start counting the other way.

Answer on page 205

8

Party Puzzles and Tricks
to Show Off With

What's the best way to clear a room at a party? Just say, 'Let me show you this trick!', which is very sad, as I love tricks and puzzles.

So, when I am sitting at a table with others and the conversation lags, if I have the props around me, I just do the trick – any trick. It's amazing how often people say, 'Wow, how did you do that?' Then it is my turn to give a little smile, which says, 'Well, it's easy if you give it some thought – which is really what your brain is for.'

Christopher Columbus was once told by some pompous twit that discovering the land that we now know as the Americas was easy. 'All you had to do was sail west until you bumped into it,' said the twit.

There happened to be a hen's egg on the table and Columbus said, 'Can you stand this egg on its end?' When the twit failed to do it, Columbus took the egg back, held it long way up and brought it down to the table with a slight tap. The tap caused the pointed end of the shell to collapse slightly and the egg stood upright. Columbus didn't actually need to say anything, as it was clear; 'Anything is easy to do, once you know how to do it.'

In this chapter I have collected puzzles and tricks for whenever a few people gather together. Mostly they are easy, once you know how. But what I want you to do is try to discover how they are done, before checking my solutions. Then you can try them on unsuspecting friends, just for the fun of it.

It's party time!

1. Get your money out

Find a fluted glass and a large and a small coin; say a 5p and a 10p. Place both coins in the glass, so the larger coin lies flat and over the smaller coin. Your impossible task: can you get the small coin out, without touching either coin?

Answer on page 205

2. The bend in the fork

Uri Geller earned himself a fortune with his magical spoon bending antics. Here is a way to fool everyone with a fork bending trick.

At the dining table, remark that the cutlery doesn't look too sturdy and pick up a fork. Hold it in both hands, as in the illustration, with the prongs resting on the table. Now, press down with both hands and the fork bends almost into a right angle. Then place the fork back on the table and, amazingly, it is totally unharmed! How?

Answer on page 206

3. Blown away

Around Christmas time, you may well find candles and wine bottles on the dining table during a meal. Here is a simple task to set your guests. How can you blow out a candle if there is a bottle in the way?

Answer on page 206

4. Candle power

How can you blow a candle out with an empty bottle?

Answer on page 207

5. Small change changing

Place three similar coins on a hard table top, so that they are in a row and touching. Call the coins A, B and C. How can you place C between A and B, without moving either A or B?

Answer on page 207

6. Chase the ace

From a pack of cards take 2 aces and 4 random cards. The 6 cards are placed face down on a table and shuffled around. If you turn 2 cards over, there are 3 possibilities:

There will be no ace
There will be 1 ace
There will be 2 aces.

So if you have a choice, which bet would you go for, in turning up two cards?

Which is more likely, to turn up 2 cards with no ace, or turn up 2 cards with at least 1 ace?

Answer on page 208

7. In black and red

Using a pack of cards once more, take an equal number of red and black cards. You can use the whole pack if you like, but just four black and four red will do.

Arrange them red/black/red/black all the way through. Now split the cards into two rough halves and riffle shuffle them together – a riffle shuffle is when you use your thumbs to flick and overlap the two halves of the deck and then bring them together into one.

A perfect riffle shuffle with alternate cards from each of the two halves is almost impossible. Also, the two halves were probably not equal when you cut them. In fact, now turn the cards over in your hands to show that they are mixed up – in a random black/red/red/black/red/black/black/red fashion.

Cut the cards, complete the cut and turn them face down.

Now, taking them from the top in pairs, what are the chances that each pair will be a red and a black?

It seems highly unlikely as your audience has just seen them mixed up, but a black and a red each time is a certainty. Why?

Answer on page 208

8. Suitably suited

This card trick is very similar to the previous one, but even more impressive.

First, sort a pack of cards so that the suit order is the same throughout, for example heart/club/diamond/ spade repeated. Show the audience what has been done.

Now ask someone to cut the cards. Quickly take one half and reverse their order, saying, 'I'll just reverse the order of half of them, so they must get mixed up.'

Now, riffle shuffle the two halves into each other.

Show your audience that the cards are no longer in their original suit order.

Then cut the cards, but be sure to make the cut between two cards of the same suit.

Complete the cut and ask someone to hold out their hands to receive the cards.

Deal four cards from the top and, amazingly, they will always be one card from each suit. But why?

Answer on page 209

9. Place them any place

Here is a game you can play with a set of dominos and an A5 sheet of paper (you can always fold an A4 sheet in half). The trick of winning this game is in the strategy that you adopt.

Player A places a domino on a sheet of A5. Player B then places a domino. The object of the game is to be the one who places the last possible domino within the bounds of the sheet of A5. Assuming you go second, can you use a strategy that ensures you win? Can you also ensure a win if you go first?

Answer on page 209

10. A good deal of dealing

Four people are playing bridge. The dealer is halfway through the deal when the phone rings.

'I must get that,' he says 'I'm expecting an urgent call.'

He puts those cards which he has still not dealt face down on the table and goes to the phone. When he comes back, he has forgotten where he was in the deal. The other three are talking and haven't touched their cards.

The dealer could count the cards in each player's pile, but there is another, quicker way of completing the deal, even though he doesn't know where the next card was to be dealt. What is that quick and certain way?

Answer on page 210

11. Making a pile

You have a pile of three ordinary dice, stacked on top of each other. You can see the faces on the four sides and the top face, but you can't see the bottom face and the touching faces. How can you quickly announce the total of the bottom side and the four touching sides?

Answer on page 210

12. You can bet it's best not to bet

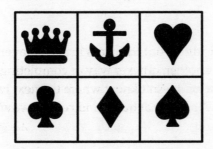

There is a very old gambling game called Chuck-a-Luck or Crown and Anchor.

The playing board had six sectors each labelled –
crowns, anchors, hearts, clubs, diamonds and spades.
The game was played with three dice with these
symbols on the faces. The dice were often kept in a
cage so they could not be tampered with. The cage was
turned upside down for each throw.

Players could bet on any of the six sectors. Then the
three dice would be rolled. Let's assume you placed a
£1 on the crown. If 1 crown showed, you would get £2
back – your pound and a winning pound. If two crowns
showed you would get £3, including £2 winnings. If the
roll showed 3 crowns, they would pay £4 – your stake
and £3 winnings.

But are these fair rewards for this gamble?

Answer on page 211

13. A toss-up between good and evil

The object of this next trick is to show how easy it is to
cheat, if that is the intention. It is not to teach you how
to cheat. So, can I trust you? Please say, 'Yes.'

Most people ask for a call before a coin is tossed or when
the coin is in the air. They then toss the coin, catch it and
slap it down on the back of their hand. Now, how could
you cheat when doing that and win the toss every single
time? Actually, it is really quite easy.

Answer on page 212

14. All the fraud of the fair

Fairgrounds have changed over the years. When I was a kid there were always coin rolling stalls. You rolled a penny and if it landed in a square without touching a line, you won the number of coins indicated on the square. Now, the squares were 1.5 times wider than the coins. So what were the actual odds of a coin landing in a square without touching a line?

Answer on page 213

15. Every one's a winner

This card trick requires a bit of preparation, but it is wonderful for a party audience. You will need two packs of cards, one with black backs and one with red backs, but you only need six cards in total. From the red-backed pack, take just the 2 of clubs. From the black-backed pack take the ace, 3, 4 and 6 of clubs and the 5 of hearts. Place them in a row, as shown in the diagram, so that all they can see is black back/ace of clubs/black back/4 of clubs/black back/2 of clubs.

Reverse of cards, not seen by the audience:

Now, from thin air you produce an 'invisible die'. It has the numbers 1, 2, 3, 4, 5 and 6 on its faces, like a normal die, but it is invisible, or to be more exact, not there at all.

In fact, only the person you choose from the audience can actually see it. You tell the person chosen that they are incredibly clever and psychic. Not only are they the only person who can see the die, but they are also the only person who will know how to find 'the phantom red card'.

So, they roll the invisible dice and tell you the invisible number on the top. What happens next?

Answer on page 213

16. A certain solution

Secretly write the number 37 on a slip of paper and place it in your pocket. Now ask someone to write three identical digits. Now ask them to divide the three-digit number by three times the digit used. As an example, if they chose 5, they would have written:

$$555 \div 5 + 5 + 5 = 15$$
$$555 \div 15 = 37$$

Whichever original digit they choose, they end up with 37. But why?

Answer on page 214

17. Christmas past, Christmas presents

On the first day of Christmas my true love sent to me,
a partridge in a pear tree. That was 1 present, but on
the second day of Christmas, my true love sent to me,
2 turtle doves *and* a partridge in a pear tree. That was
3 more presents – 4 in total. We know, of course, that this
continued for the entire 12 days of Christmas.

How many presents arrived on the 12th day of
Christmas? How many presents arrived over the entire
12 days?

Answer on page 214

18. Gone to pot

Most people know from watching snooker on television
that a red counts 1, a yellow counts 2, a green counts –
never mind, you don't need to know any more. A snooker
player goes to the table and makes a break of just 8, in
which he pots 4 yellow balls. Can you explain how?

Answer on page 215

19. A mixture mix-up

You have a glass half full of wine and another half
full of water. You take a tablespoon full of wine and
drop it into the water. You now take a tablespoonful of
this mixture and drop it into the wine. Now, does the
mostly-wine contain more water than the mostly-water
contains wine, or is it the other way around? Or is the
amount of wine and water in each, exactly the same?

Answer on page 215

20. Card sharp

From a pack of cards, take ten cards from ace to 10.
The suits do not matter. Now, draw a pentagon on a
piece of paper and place the cards at the points and the
mid-points of the lines so that all five sides composed
of three cards in a row, add to the same number. It is not
too difficult.

Answer on page 216

21. Party pooper

At a Christmas party, you often get cone-shaped hats,
or alternatively you can make one to use in this trick in
seconds.

Cut the tip off a cone-shaped hat to form a small hole.
Place a lit candle about 15 cm from the wide end of the

cone. Now, by blowing through the hole and down the cone, is it possible to blow out the candle?

Answer on page 216

22. Counter-productive

This is a puzzle you can set for someone when you have a set of backgammon counters handy.

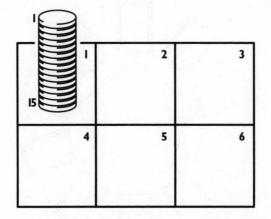

Quickly draw a set of 6 squares on a piece of paper and number them 1 to 6. Using a pencil, lightly number each counter from 1 to 15 and place them in a single pile on square 1 in number order, 1 on top and 15 on the bottom.

The object is to move them one at a time onto other squares until the whole pile is transferred to square 6, using the other squares along the way.

But no counter must ever be placed on a counter with a smaller number.

Answer on page 217

Answers

1. Kitchen Capers and Domestic Problems

1. Bun fun with Mum

Yes, you can. First divide the buns into 3 sets of 9.

Weigh 9 against 9. Whichever side goes down contains the heavy bun. If they are equal then the heavy bun is in the 9 you haven't weighed.

For the second weigh, divide the remaining 9 into 3 sets of 3. Weigh a set of 3 against another set of 3. You will thus find the set that contains the heavy bun, in the same way that you found the correct nine in the step above.

Now, from the last 3 weigh one bun against another. From this weigh you will know precisely which is the odd cross bun with the £2 coin inside.

2. The sands of time

To time 15 minutes exactly, start both sand timers running.

When the 7-minute timer has emptied, turn it over.

When the 11-minute timer has emptied, turn the 7-minute over again.

The 7-minute glass will have just run 4 minutes since you last turned it, so it will take 4 minutes to run back again – finish on 15 minutes exactly.

3. One green bottle

You have to get round the fact that the top end of the bottle is tapered.

You need to find the height of the inside of the bottle if it were a smooth cylinder. Here's how you do it.

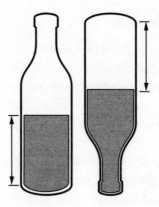

Measure the depth of the liquid. Now turn the bottle upside down. Make sure the cork is in first, dummy!

While it is upside down, measure the height of the space above the liquid.

Add the two measurements together.

Multiply by the 30 cm² cross sectional area and you have the internal volume.

4. A toast to toast

We'll call our 3 slices of bread A, B and C.

First toast A and B one side.

Now toast A's other side and C on its first side.

Finally toast both B and C on their second side, and you have toasted 3 pieces in 3 toastings. You should be the toast of the town! Or at least the kitchen.

5. Time on his hands

He wound his clock and set it at 12.00 and left to walk to
  his mate's house.
When he arrived, he noted the time immediately.
He stayed and chatted.
He noted the time as he left and walked home at the
  same pace.
Back home, he deducted the time he had stayed with his
  friend from the time shown on his clock.
He then divided the time since 12.00 by 2.
This was the exact time he had taken to walk home.
He added the time of the walk home to the time on his
  friend's clock when he left.
Now he knew the correct time – and he could now set
  his own clock right. Right?

6. Draw your own conclusions

There are 37 pencils in the hexagonal bundle. There are a couple of ways to work this out.

EITHER

1 pencil can be surrounded by 6 pencils. You can surround 6 pencils with 12 and still keep the hexagon shape. You can then surround the 12 with 18 and still keep the hexagon shape. So, 1 + 6 + 12 +18 = 37

OR

There are 18 pencils around the bundle. So each side of the hexagon has 3 pencils plus an end one. So there must be a row of 4, then one of 5, then one of 6, then one of 7. And then diminishing rows of 6, then 5, then 4. So, 4 + 5 + 6 + 7 + 6 + 5 + 4 = 37.

7. Sock it and see

First place the 2 red socks and imagine 3 gaps:

R – – – R –.

Then add the 2 blue socks with 1 gap:

R B – B R –.

Now add the greens with 2 gaps.

R B G B R G.

Sock-cess!

8. Put a sock in it

You have 8 sock spaces.

Starting with the yellows, which need 4 spaces between, you have only 2 options:

– Y – – – – Y –

OR

Y – – – – Y – –.

For the first there are only two reciprocal ways to add the reds, like this:

R Y – – R – Y –.

There is now no way to add the greens and blues according to the rules.

So you must start:

Y – – – – Y – –.

Try adding the reds like so:

Y – – R – Y – R.

Now it is impossible to add the greens and blues according to the rules!

So you must place the reds like this: Y – R – – Y R –.
Now you can add blue and green
to give: Y B R B G Y R G.

Another feat accomplished!

9. A sucker born every minute

Forty minutes is the answer. You start sucking a sweet.
You will finish it and pop the second one in your mouth
after 10 minutes, the third after 20 minutes, etc. That
means that the fifth will disappear into your gob after
40 minutes and you will have none left, though you will
be sucking the last one for another 10 minutes.

10. An eggs-acting question

To solve this problem you must work backwards.

Daisy stole half the eggs plus half an egg and left no
eggs at all. So Daisy must have stolen ½ + ½ = 1 egg.

Celia bought half the eggs plus half an egg and left
1 egg. So from 3 eggs she bought 1½ + ½ = 2

Betty did the same, so from 7 eggs she bought 3½ + ½ =
4 and left 3.

Anne must have seen 15 eggs. She bought 7½ + ½ = 8
and left 7.

So, successive customers take eggs in multiples of 2. In our case they took 8, 4, 2 and 1 egg respectively from the original 15 that Mary started with.

11. Put your money where your mouth is

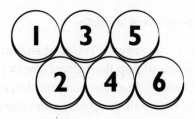

Did you find the correct first move?

Move 2 to touch 4 and 6

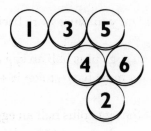

Now move 4 to where 2 was, to touch 1 and 3

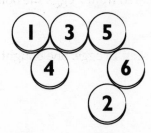

Lastly move 1 to touch 4 and 2 and complete the circle.

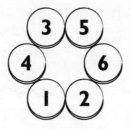

There are actually two possibilities, as the puzzle is reciprocal. So, you could start by moving 5 to touch 1 and 3.

12. Fumbling in the dark

I would need to take 4 shoes to be sure of having a pair that match, including a right and a left of the same pair.

I would only need to remove 3 socks to be sure to have a matching pair. Mind you, the socks might not match the shoes … but these days that doesn't seem to matter. However, if you must have, say, black socks, then you need to take 14 to be certain of having a pair.

13. Animal farm

He would still have 50 horses – calling a cow a horse doesn't make it a horse.

14. The weight of the matter

The weights need to be in unit weights of 1, 3, 9, 27 and 81. Notice each new weight is 3 times the previous one. With just these 5 and a balance scale, you can weigh any unit weight up to 121.

As examples:

To weigh 2 units, put the 3 on one scale and the 1 on the other. The difference is 2.

To weigh 74, place 81 and 3 on one side and 1 and 9 on the other, $(81 + 3) - (9 + 1) = 74$.

The next higher weight would need to be 243. With these 6 weights you could then weigh every unit weight up to 364.

Compare that with six doubling unit weights, as used in the days before metric – 1, 2, 4, 8, 16, and 32. With these you can only weigh units up to 63.

But with the modern metric system you would need weights for 1, 1, 2, 5, 10, 20, in order to measure all unit weights from 1. Sadly this set will only allow you to weigh all units up to 39. So much for progress!

15. You could lose your balance

The secret is to realise that you can swing the balance scales around.

So, from (1) we know that: a bottle and a glass = a jug.

And from (2) swapped around we find that: a glass and a plate = a bottle.

Add the two together and you get: a bottle, 2 glasses and a plate = a bottle and a jug.

Remove the bottles from each side and you get: 2 glasses and a plate = a jug.

That means that: 4 glasses and 2 plates = 2 jugs.

But from (3) swung around we know that: 3 plates = 2 jugs.

So: 4 glasses and 2 plates = 3 plates.

So: 4 glasses = a plate.

Now, from (2) we know that: a bottle = a glass and a plate.

So, finally, a bottle must = 5 glasses.

Have your eyes glazed over?

16. A pond to ponder over

The answer is not 15 days. As they double in size each day, but this year there are two of them, we can assume they would cover a pond twice as big in 30 days. So they will cover the original pond in 29 days and not before.

17. Early doors

There is a simple way to explain all 'think of a number' puzzles. Re-examine it, without the numbers you are asked to put in. The sequence goes: think of a two-digit number, double it, add 5, multiply by 50.

That is the same as multiplying the two-digit number by 100 and adding 250.

Then you add 365 to the 250 = 615. Take 615 away again.

So the two-digit number is now multiplied by 1000 or moved two places to the left.

This leaves space for your age to follow.

18. Measure for measure

There are two ways of doing this, one slightly quicker than the other.

	8	0	0
Starting position …	8	0	0

Solution 1

From the 8-unit jug, fill the 5-unit jug.	3	5	0
From the 5-unit jug, now fill the 3-unit jug.	3	2	3
Empty the 3-unit jug back into the 8-unit jug.	6	2	0
Pour the 2 units left in the 5-unit jug into the 3-unit jug.	6	0	2
Fill the 5-unit jug from the 8-unit jug.	1	5	2
Pour 1 unit from the 5-unit jug to fill the 3-unit jug.	1	4	3
Empty the 3-unit jug into the 8-unit jug.	4	4	0

Now there are 4 units in the 8- and 5-unit jugs.

Solution 2

From the 8-unit jug fill the 3-unit jug.	5	0	3
Pour that into the 5-unit jug.	5	3	0
From the 8-unit jug now refill the 3-unit jug.	2	3	3
From the 3-unit jug, fill the 5-unit jug, leaving 1 unit.	2	5	1
Empty the 5-unit jug into the 8-unit jug.	7	0	1
Pour the 1 unit from the 3-unit jug into the 5-unit jug.	7	1	0
Now fill the 3-unit jug from the 8-unit jug.	4	1	3
Pour that into the 5-unit jug.	4	4	0

You now have 4 units in the 8- and 5-unit jugs.

The first method uses 7 moves, but the second requires 8 moves. Also, along the way you had, at one time or another, exactly 1 pint, 2 pints, 3 pints, 4 pints, 5 pints, 6 pints and 7 pints. In the first solution every unit except 7 was achieved. In the second solution every unit except 6 is achieved.

19. Milk of human kindness

First the milkman churned the problem over in his mind. Then he delivered exactly 2 litres to each lady, in a roundabout sort of way.

From one 10-litre churn he filled the 5-litre jug.
From that he filled the 4 litre jug, leaving 1 litre in the 5-litre jug.
He put the 4-litres back in the 10-litre can.

He then poured the remaining 1 litre from the 5-litre jug into the 4-litre jug.

From the same 10-litre churn, he filled the 5-litre jug, and from that he filled the 4-litre jug.

He was left with exactly 2 litres in the 5-litre jug.

That was one lady served – she paid and popped back indoors. But what about the second lady, with her full 4-litre jug?

The milkman emptied the 4 litres back into the original churn, so that it contained 8 litres.

From the churn so far not used, he filled the 4-litre jug.

He then topped up the original churn with 2 litres from the 4-litre jug.

That meant that the second lady had 2 litres and she was satisfied as well.

As they say, 'One good churn deserves another!'

20. Going for the juggler

Using a 7-pint and a 10-pint jug, here's how you measure exactly 9 pints …

Fill the 10-pint jug from the tap. From that fill the 7-pint jug, leaving 3.

Empty the 7-pint jug down the sink.

Pour the 3 pints into the 7-pint jug. Fill the 10-pint jug from the tap.

Fill the 7-pint jug from it, leaving 6 pints in the 10-pint jug.

Pour away the 7-pint jug.

Put the 6 pints into the 7-pint jug.

Fill the 10-pint jug.

Add 1 pint to the 7-pint jug, leaving 9 pints in the
10-pint jug.

Success!

21. More jug juggling

Measuring 1 unit using 7- and 9-unit jugs

Fill the 7-unit jug from the tap.

Pour the 7 units into the 9-unit jug.

Fill the 7-unit jug again.

Top up the 9-unit jug, leaving 5 units in the 7-unit jug.

Empty the 9-unit jug down the drain – good job we are
using water, not milk!

Pour the 5 units from the 7-unit jug into the 9-unit jug.

Fill the 7-unit jug.

Top up the 9-unit jug, leaving 3 units in the 7-unit jug.

Empty the 9-unit jug down the drain.

Pour the 3 units from the 7-unit jug into the 9-unit jug.

Fill the 7-unit jug.

To the 3 units in the 9-unit jug add 6 units to fill it.

This will leave you with just 1 unit in the 7-unit jug – job
done.

If you fill the 7-unit jug first, the operation requires
12 moves and wastes 18 units of water. Filling the 9-unit
jug first takes 16 moves and wastes 28 units of water.

Measuring 1 unit using 7- and 12-unit jugs
I am not giving the answers in detail here, but of the
two possible solutions filling the 12-unit jug first is best,
requiring 14 moves. Filling the 7-unit jug first requires
20.

Measuring 1 unit using 8- and 13-unit jugs
In this case it is best to fill the 8-unit jug first – the
operation then requires 14 moves. Filling the 13-unit jug
first means that it requires 24 moves.

22. Firing on two cylinders

The secret is in the fact that the jugs are cylinder shaped.

Carefully start to pour from the 8-pint cylinder into the
5-pint cylinder. Keep pouring, and slowly tilt the 8-pint
cylinder. Watch the level as you tilt. When the level of
the milk (or water) is in line with the upper edge of the
base circle, and with the pouring lip at the other end, the
cylinder is exactly half full. Stop pouring – job done.

2. 'I Can Hear You Thinking' Puzzles

1. On the right lines

The locomotive moves to B, then hooks up with X and
takes it via B to C.

He then pushes goods van Y to A, moves around via B
and pulls goods van Y to B, then across to C, to link up
with goods van X.

Now he pulls them both across to B, then shunts Y and X
up to A.

He disconnects and leaves Y in the original position
of X.

He then goes via B and C to pull X to position Y.

2. Between flights

It sounds complicated, as it seems that you would have
to work out the time for each journey the fly takes, as
they get shorter and shorter. But it is much easier than
that.

The lads are 20 miles apart and each cycle at 10 mph,
covering the distance in 1 hour.

The fly is flying at 15 mph. So in one hour he flies
15 miles.

It's as simple as that.

A famous Hungarian mathematician, John von Neumann,
was asked this question and gave the answer very
quickly. The questioner said, 'Some people get tricked
and try to work it out by finding the sum of the infinitely

reducing series of journeys the fly takes.' Von Neumann said, 'Yes, that is the system I used!' Wells.

3. It's a knockout

Every match provides 1 winner and 1 loser who is then out of the running . If there are 29 teams at the start, you must need 28 matches to knock out all but the winner.

4. The Inn of the Sixth Happiness?

It all sounds too good to be true, and it is. In placing the lads in the rooms the manager never placed the 2nd lad anywhere.

5. Talking balloons

As the car swung left, there would be a slowing in the forward direction, as well as the move left.

So, he and his son and anything else heavy and loose in the car would move forward and right – let's hope they were wearing their seat belts.

But the balloon filled with helium would be lighter than the surrounding air. So, the heavier air moving forward and right would push the balloon backwards and left.

6. Between the sheets

As the first page is page 1, each sheet must have the odd number on the first side and the next highest even number on the other side.

So, if page 6 is missing, then 5 is also missing. If 19 is missing, then 20 is also missing. Also, as there are 4 pages before the missing pages, there must be 4 after them, so the total number of original pages must have been 24.

7. Old before his time

The chap must have been born on the 31st December and must have been talking to me on the 1st Jan. That would mean on the 30th December he was 19, and on the 31st, he was 20. So on the 31st of this year he will be 21 and on the last day of the following year he will be 22.

8. Flights of fancy

There are two equally correct answers.

Either he could have been intending to fly right around the world or he could have been heading for a spot on the globe exactly half way around the world from their current location.

In both cases he could head for the girl's destination, drop her off, and then carry straight on in the same direction, without going out of his way.

9. What to wear where

Miss Black cannot be wearing green, as the girl in green answered her.

As Miss Black cannot be wearing black either, she must be wearing white.

The girl wearing green cannot be Miss Black, so she is Miss White, and if Miss White is wearing green, then Miss Green is wearing black.

10. The floating hat

This puzzle sounds complicated, but it's not.

As all the action takes place on the river, the river's speed of flow can be discounted. He first rows upstream away from his hat. Then on turning and rowing at the same stroke and rowing speed, he will take the same time to reach his hat – 3 minutes.

Only to those on the bank, will he have appeared to be going slower upstream and faster downstream. His speed will always be the same relative to the water.

11. Truth and Liars Club

The truth-teller is the man who said there were 8 around the table. For the first statement to work there would have to be an even number of people around the table and they would have to be alternately truth-tellers and liars.

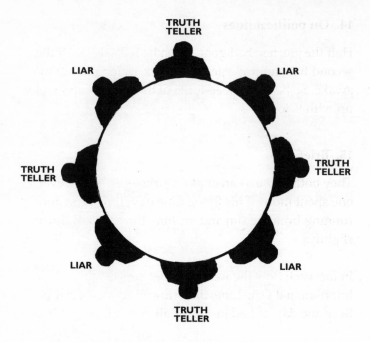

12. Smitten by kittens

She has ¾ of their number plus ¾ of a cat.

So ¾ of a cat must be ¼ of all the cats.

So she has 4 times ¾ cats or ¹²⁄₄, which equals 3 cats.

13. Suitably attired

He spent twice as long walking ⅓ of the way, as he spent riding ⅔ of the way. So he cycled 4 times as fast as he walked.

14. On political lines

Half the journey had gone when he fell asleep. Of the second half, he was asleep for twice as long as he was awake. So he was asleep for ⅓ of the journey and woke up with ⅙ to go.

15. Brief lives

They both chose a career as an airline pilot. However, one spent more of his flying time travelling west and running from the sun and for him, time slowed down slightly.

In fact each time the west-traveller crossed the International Date Line, time slowed so much that he lived the day he had just lived all over again.

The other brother spent more of his time travelling east and for him, time speeded up. When he crossed the International Date Line the other way he lost a day and time speeded up even more dramatically.

So, unless they both crossed the International Date Line exactly the same number of times each way, the one travelling west more would have slowed time down and lived longer than the one travelling more to the east.

16. Merrily we roll along

It is worth considering three different points as the rollers progress – the point touching the ground, the centre of and the top surface.

On a rolling roller, or any wheel, the point touching the ground does not move at all. On any wheel, the centre hub always moves forward at the same speed as the vehicle.

But the slab in our question is only in contact with the top edge of the rollers, and the top edge of a wheel is always travelling at twice the forward speed of the wheel itself.

So, in just one roller revolution, each roller will have moved 3½ metres but the slab will have moved twice that, or 6⅔ metres. Already three of the rollers will no longer be under the slab and will need to be moved forward.

17. It's in the bag

You might think the answer would be ⁵⁰⁄₅₀, either black or white. But you would be wrong.

If the original counter was black, then the one left would certainly be black.

But if it was white, the counter taken out could be either of the two white counters. So the counter left inside must be either black or one of the two white counters.

So the answer is that the probability of the second counter being white is ⅔.

18. The vampire umpire

At the start, the vampires must have the boat on their side.

First 1 vampire only is told to come across.

Next 2 tourists cross, leaving the guide and 1 vampire behind.

Now 1 tourist and 1 vampire come back across.

The tourist and guide now complete their crossing.

Finally, the last vampire rows across and not a drop of blood has been spilt.

19. The prisoner's get-out clause

The prisoner's dilemma is that he can't visit 9 cells and arrive at the exit without doubling back on himself.

Whichever route he takes, he cannot visit each cell once and once only.

The secret is in the instruction 'enter each and every cell once only'.

He must first exit his own cell and enter an adjacent cell.

Then he must enter his own original cell.

Now it is simplicity itself to visit all the other cells and arrive at the exit door.

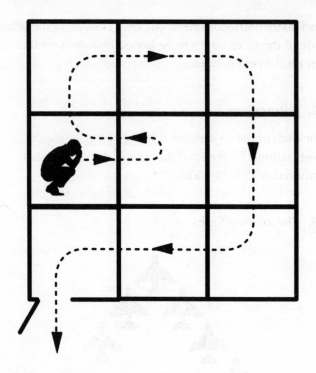

20. Backward drawbacks

I was travelling at 55 mph.

In 2 hours I must have travelled 110 miles, because 15951 + 110 = 16061, which is the next possible palindromic number.

21. The answer to four across?

The kids across first; one comes back; Dad rows across; the other kid comes back; now both kids cross; one comes back; Mum rows across; the other kid comes back and lastly both kids cross. The boat is now on the other side of the river, so it's to be hoped that another family isn't following behind.

22. Journey's end

For each of the 15 stations there are 14 possible other destinations. So the total number of different tickets required is $15 \times 14 = 210$.

23. The arrow's flight

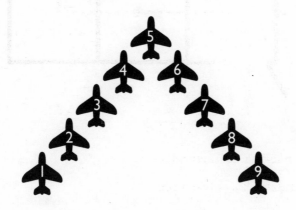

He couldn't have been in an end plane, as he had to have planes each side of him.

Assuming he was in position 2, he would have 1 plane left and 7 right of him: $1 \times 7 = 7$.

Assuming position 3, he'd have 2 planes to his left and 6 to his right: $2 \times 6 = 12$.

At position 4, he'd have 3 planes left of him and 5 planes right of him: $3 \times 5 = 15$.

At position 5, in the centre, he'd have 4 to both his left and right: $4 \times 4 = 16$.

At position 6, there would be 5 to the left and 3 to the right: $5 \times 3 = 15$.

At 7, 6 left and 2 right: $6 \times 2 = 12$.

At 8, 7 left and 1 right: $7 \times 1 = 7$.

Positions 3 and 7 both give $2 \times 6 = 12$, 3 less than 4 and 6 which give 15.

He couldn't possibly be 3 right of position 7.

So he had to be in position 3.

24. Soul sisters

Obeying the rules that no woman is ever left with a man when her brother isn't present, it can be done in 11 crossings, and not less. If you say you have done it in 9 crossings, check, because one of the men will have been with an unprotected sister, while one is getting into or out of the boat.

Here is the correct answer as told by one of the men …

I cross with my sister, leave her and bring the boat back.
The other 2 girls cross and my sister comes back and
   I help her out of the boat.
The other 2 fellas cross, making 2 couples on the other
   side.
One couple comes back.
The chap and I cross over, leaving the 2 sisters.
The third sister comes back.
Two sisters come over.
Lastly one man goes back to bring his own sister across.

25. Driven to distraction

The first driver takes 2 hours 20 minutes to go and
1 hour 45 minutes to come back – 4 hours 5 minutes.
The second driver takes 4 hours exactly and so saves
5 minutes. All that rush for 5 minutes? Tut tut.

26. Puffed out

Fred decides to smoke ⅔ of a cigarette each day.
After 9 days he has left 9 stubs, each ⅓ of a cigarette
   long.
Using cigarette paper he joins the stubs together to make
   3 more cigarettes.
He smokes ⅔ of a cigarette for the next 3 days and now
   has another 3 stubs left.
They make 1 last cigarette, which he smokes on the
   13th day.

He has finished smoking within the 2 weeks as
promised.

27. Double expresso no so fasto

That trains can get past each other like this …

The right engine (RE) unhooks and pops into the siding.
The left train passes the siding and stops.
The right engine passes to the left behind the left train.
The left train picks up carriage RC1 at the right and
    shunts it into the siding.
The left train moves off to the right.
The right engine picks up RC1 from siding.
The left train puts carriage RC2 into siding and then
    goes on its way.
The right engine picks up RC2 and goes on its way.

---

3. Any Number of Puzzles – About Numbers

1. Puzzle this ONE out …

A set of six 1s squared gives $111,111 \times 111,111 =$
$12,345,654,321$.

A set of seven 1s squared gives $1,111,111 \times 1,111,111 =$
$1,234,567,654,321$.

Squaring number of 1s (squaring means multiplying a
number by itself) gives a beautiful pattern of answers …

$$1 \times 1 = 1$$
$$11 \times 11 = 121$$
$$111 \times 111 = 12,321$$

And eventually, $111,111,111 \times 111,111,111 =$
$12,345,678,987,654,321$.

2. Eight missing

Because $18 = 2 \times 9$, $12,345,679 \times 18 = 222,222,222$.
Because $45 = 5 \times 9$, $12,345,679 \times 45 = 555,555,555$.
Because $81 = 9 \times 9$, $12,345,679 \times 81 = 999,999,999$.

3. Looking both ways

$123,456,789 + 987,654,321 + 1 = 1,111,111,111$.

So $123,456,789 + 123,456,789 + 987,654,321 + 987,654,321$
$+ 2 = 2,222,222,222$

4. Sum for simpletons

$56 = 7 \times 8$.

5. In seventh heaven

If you divide $999,999$ by 7, the answer is the magic
number $142,857$.

Now, $142,857 \times 2 = 285,714$.
Also, $285,714 \times 2 = 571,428$.
And, $428,571 \times 2 = 857,142$.

In each case the second number is obtained by sliding
the first 2 digits to the rear end.

In fact all 6 numbers are versions of $142,857$. Each starts
with a different digit but the digits are always in the
same order.

The 6 numbers in order are $\frac{1}{7}$, $\frac{2}{7}$, $\frac{3}{7}$, $\frac{4}{7}$, $\frac{5}{7}$ and $\frac{6}{7}$ of
999,999.
By dividing 1,000,000 by 7, you get 142,857.142,857
recurring.
Try dividing 1,000,000 by $\frac{2}{7}$, $\frac{3}{7}$, $\frac{4}{7}$, $\frac{5}{7}$ and $\frac{6}{7}$.

Many ingenious maths tricks have been designed using
the magical 142,857. Notice that 142,857 includes all the
single digits except those divisible by 3 – 3, 6, 9.

Here is a magical way of writing 142,857 if you have
forgotten it:

Write 6 dashes	---,---.
In the first place write the 1	1--,---.
From the 1, count 2 places and write the 2	1-2,---.
Now count 4 dashes and place the 4	142,---.
Count 5 dashes and place the 5	142,-5-.
Count 7 dashes and place the 7	142,-57.
Place the 8 in the final place to get	142,857.

6. Pure Gauss work

To add all the numbers from 1 to 100, quickly, there
has to be some quick method. As there are only two
numbers mentioned in the question, perhaps they will
lead you to the answer?

What do you get if you add 1 + 100? 1 + 100 = 101.
Now, take the two numbers one place in from the 1
and the 100, that's 2 + 99 = 101.
Then, what a surprise, 3 + 98 = 101.
And so it continues …

So how many pairs of number can you make from 100 numbers? 50.

So the answer is $1 + 100 = 101 \times 50 = 5050$. So our song went,

> *Carl Freidrich Gauss when only nine, he did a sum in record time,*
> *'Please do this sum,' his teacher said, 'add the numbers from one up to a hundred.'*
> *He made a pair of first and last, that's one hundred and one,*
> *And multiplied by fifty pairs and there the job is done.*
> *It's one of the tales of maths and legends, one of the tales of maths and legends.*

In fact, the sum that Gauss did when he was nine years old was a little more complex. His teachers asked, 'Starting with 48,617, then 48,634, then 48,651, what is happening?'

The answer is that each new number is 17 higher than the previous number.

'OK,' said the teacher, 'give me the total of the first 100 terms in this sequence, starting with 48,617.' This is a much more complex puzzle than adding 100 consecutive numbers. Gauss wrote the answer down on his slate – this was in the days when paper was too expensive to waste on school work. He showed no 'working out', as he had done all that in his head.

Gauss was extremely gifted and became one of the greatest mathematicians of all time. However, the fact that he was asked this question aged 9, shows how much further advanced his teaching was, compared with the maths taught to 9-year-olds today. I believe there are thousands of gifted 9-year-olds out there, but that our system today is failing to find many of them and identify their special abilities. This is a tragedy not just for them, but for everyone.

7. Premier division

What number is divisible by all the digits from 1 to 9, leaving no remainder? Well, you can reduce the numbers down from 9:

All numbers are divisible by 1, so forget that.
If a number is divisible by 9, it is also divisible by 3, so
  you don't need the 3.
If it is divisible by 8, it also divides by 4 and 2, so you
  don't need them.
If it divisible by 9 and 8, it must be even and thus
  divisible by 6, so that is not needed either.
So you are left with just 9, 8, 7 and 5.
So, $9 \times 8 \times 7 \times 5 = 2520$, and that is the answer.

8. An imaginary menagerie

Let's start with the eyes.

If there are 60 eyes then there must be 30 animals in total.

Now, there are 86 feet altogether, and we know that every animal has at least 2 feet, so that's 60 feet.

So, 86 − 60 = 26, which is the number of extra feet, belonging to camels!

Which means there are 26 ÷ 2 = 13 camels.

If there are 30 animals altogether, then 30 − 13 = 17 ostriches!

9. Starter for ten

If Alf starts 10 metres further back, when Beryl has run 90 metres, Alf will have run 100 metres.

So they will be level.

But then Alf will run the last 10 metres while Beryl runs 9 metres.

So Alf will win again.

In order for them to draw, Alf must start at the 100 metres mark and give Beryl a 10-metre start.

10. A Greek bearing birthday gifts

Sausagyknees was 69. There was no 0 AD or BC, so he was 34 in 1 BC and 35 in 1 AD. So on his birthday in 35 AD, he was 69. To be nit-picky, if he died before his actual time of birth, he was still only 68.

11. Withdrawal symptoms

There is no missing extra pound – the right hand balance column isn't supposed to add up to £50. If you took out £1 a day for 3 days, the left column would be 1 + 1 + 1 + 47 = 50. But the balance column would show 49 + 48 + 47 = 144, which is ridiculous. Only the left-hand column will give you a true total of your withdrawals.

12. In proportion

You must add 20 to each. 100 + 20 = 120 and 20 + 20 = 40. $^{120}\!/_{40}$ equals a ratio of $^3\!/_1$.

13. On the square

The numbers are 6 and 8, which are in the ratio of 3 to 4. $6 \times 6 = 36$ and $8 \times 8 = 64$.

$$36 + 64 = 100.$$

14. Spare me more squares

The two numbers are 10 and 30. $10 \times 10 = 100$ and $30 \times 30 = 900$. $100 + 900 = 1000$.

15. Shanks's pony

However long it takes for the first half of the journey, the horse will take 3 times as long for the second half. Let's say the journey is 24 miles. It goes at 12 mph and takes 1 hour for the first half. But at 4 mph it would take 3 hours for the second half. So it would do 24 miles in 4 hours, which is 6 mph.

You cannot add the two speeds together and then half the result ($12 + 4 = 16$ and $16 \div 2 = 8$) as the horse travels for a far longer time at the slower speed.

16. Has the penny dropped?

You shake the 15p box to glimpse one coin.

This box cannot contain 15p, as all the labels are incorrect, so on seeing one coin, you will know that the other coin is exactly the same. Let us assume you see a 5p, in that case the box contains two 5ps.

Now, the two 10ps cannot be in the 20p box so they must be in the 10p box. As a result you know that the 20p box contains 15p made up of a 5p and a 10p coin.

Let's assume you see a 10p first instead. If you do then the other coin must also be a 10p. Now the two 5ps cannot be in the 10p box, so they must be in the 20p box. That means that you know that the 10p box contains 15p made up of a 5p and a 10p coin.

17. Dozen really matter

Six dozen dozen is $6 \times 12 \times 12$, or 864. Half a dozen is six. So half a dozen dozen is $6 \times 12 = 72$. So, the difference is $864 - 72 = 792$. But in dozens, the answer works out at five and a half dozen dozen. Dozen it?

18. It's people that count

B will always have exactly 20 counters. But did you work out why? Let's go through it once more.

Let us assume A takes 7 counters to start with.

B now takes 3 × 7 = 21 counters.

A now gives B 5 of his counters, leaving A with 2 and B with 26.

B now gives A 3 × 2 = 6.

As he had 26, he is now left with 20.

A will always end up with 4 times the number he started with, greater than 5. B will always end up 3 × 5, plus another 5 = 20, having given all other counters to A.

19. The missing numbers

This is not too difficult.

The number after the 5 must be 7, as 7 × 9 (63) ends in 3. So, 57 × 9 = 513.

The number after 9 in the answer must be 5 and 57 × 5 = 285.

The third number must be 3 and 57 × 3 = 171.

It is now quite easy to fill in the missing digits and finish the puzzle.

$$
\begin{array}{r}
953 \\
57\overline{)54321} \\
513 \\
\hline
302 \\
285 \\
\hline
171 \\
171 \\
\hline
0
\end{array}
$$

20. Never ask a lady her age

To have added a half, a third and 9 (i.e. 3 × 3), the age
  must have just about doubled.
Also, to be able to add a half and a third, the age must
  be divisible by 6.
'Six score and ten' is 130 and half of that is 65.
As the answer is divisible by 6, the answer cannot be 65.

Let's try 60:

$$60 + 30 \, (\tfrac{1}{2}) + 20 \, (\tfrac{1}{3}) + 9 = 119$$

No, the answer we're looking for is 130.

So let's try 66:

$$66 + 33 + 22 + 9 = 130.$$

So the answer is, 'She be 66 years old.'

21. A calculated guess

Keying in a three-digit number twice to form a six-digit
number is exactly the same as multiplying the three-
digit number by 1001. If we multiply the divisors we
get 7 × 11 × 13 = 1001. So the exercise of dividing the
original six-digit number by these numbers, in any
order, will always leave no remainder and bring us back
to the original three-digit number.

22. Triple tapping

If you write any two-digit number 3 consecutive times, it is just the same as multiplying the original number by 10,101.

But $3 \times 7 \times 13 \times 37$ also equals 10,101.

23. Gifted sons

There are only three people involved, son, father and grandfather.
The grandfather gave his son, the father, £100.
He then passed on £50 to his son, the grandson.

24. Think of a number

18 equals twice its digits.
27 equals three times its digits.
36 and 48 both equal four times their digits.

25. Chinese crackers

In each case, make a list of numbers.

Numbers that divide by 3 with remainder 2:

$$5, 8, 11, 14, 17, 20, 23, 26, 29, 32, 35.$$

Numbers that divide by 5 with the remainder 3:

$$8, 13, 18, 23, 28, 33.$$

And numbers that divide by 7 with the remainder 2:

$$9, 16, 23, 30, 37$$

The only number that appears in all 3 sequences is the number 23.

26. Three little maids – in a flat

To find the answer, you have to make a table of when each girl will be home.

The 5-day girl will be home on days 5, 10, 15, 20, 25, 30, 35, 40, 45, 50, 55, 60.

The 4-day girl will be home on days 4, 8, 12 16, 20, 24, 28, 32, 36, 40, 44, 48, 52, 56, 60.

The 3-day girl will be home on days 3, 6, 9, 12, 15, 18, 21, 24, 27, 30, 33, 36, 39, 42, 45, 48, 51, 54, 57, 60.

The first day that coincides, when all three girls will be home together again, is day 60.

But there was actually an easier way to find that day. Since 5, 4 and 3 have no similar factors, simply multiply $5 \times 4 \times 3 = 60$.

4. Easy Peasy Puzzles and Catchy Watchy Questions

1. Off the record

On a modern computer hard disc, there can be 143,000 parallel tracks in each radius inch or 2.5 cm. But on a vinyl record, how many grooves? Just two, one on each side.

2. Look at it another way

80 minutes and 1 hour 20 minutes are exactly the same time.

3. Stock is money

They are door numbers from a hardware store.

4. Crows' feat

The 3 dead ones – the rest will have flown away!

5. You have the gift

You will have 3 apples left.

6. Uncorking a corker

The bottle costs 19.5p and the cork 0.5p.

7. Eggsasperation

He kept ducks.

8. Cutting it fine

Because he will make twice as much money!

9. A catfish conundrum

If a cat and a half eat a fish and a half in a day and a half, then a single cat would eat a fish in a day and a half.

So a single cat would eat 7 fish in a week and a half.

Seven cats would eat 49 fish in a week and a half.

10. Ark at this

None! It was Noah who built the Ark, not Moses.

11. Tweedling their sums

It is clear that Tweedledee is an odd weight and Tweedledum is an even weight. So, Tweedledum weighs 50 kg and Tweedledee weighs 51 kg.

12. A word to the wise

The right answer is: 'wrong'. And any other answer is also wrong!

13. Under a cloud

There is no possible chance of sunshine – 72 hours is
3 days exactly, and therefore it will still be midnight.

14. Private lives

The colonel was the mother of the private.

15. Have a break, have a ...

If it's a 20-segment bar then the answer is 19. Each time
you snap a piece of chocolate, the total number of pieces
increases by 1. You could stack and tear paper and so
reduce the number of tears, but you can't stack chocolate
– can you?

16. Mind-numbing numbers

The two whole numbers are 1 and 7. Doh.

17. Only the half of it

Did you get as an answer, 33?

That is not correct, because 50 divided by ½ does not
give you 25.

$50 \times ½ = 25$. $50 \div ½ = 100$, as there are 100 ½s in 50.

So 50 divided by ½ plus 8 is: $100 + 8 = 108$.

18. Percent to get you!

No, they should not be happy. Cutting 20% and then adding 20%, doesn't get you back to the same number.

Imagine someone's wage was £400.
A 20% cut would be £80, leaving them with £320.
But if later 20% were to be added on to £320 that would be only £64.
So the newly adjusted wage would be £384.

That would mean that the employee would be £16 worse off. Crafty boss!

19. To coin a phrase

One is not a 10p, but the other one is.

20. Got you surrounded

Draw the circle around their waist, either on their skin or on the clothes they are wearing.

21. The magic money clip

Yes you can. Curve a banknote to form an 'S' shape with two loops. Place two paper clips to join each end to the nearest loop.

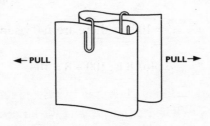

Now simply pull the two ends of the note sideways. Magically, the two clips fly off the bank note. Close examination will show that they are now linked.

22. In the cold light of day

The match.

---

5. Geometric Shape and Angle Puzzles

1. Halving the pain

Place the new window so that each corner is half way along an edge of the larger frame, like a square diamond.

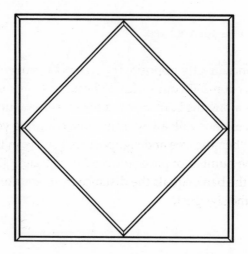

Imagine the original was made up of four squares – you have bisected and halved each of those squares. The new window will be half the area of the large square.

2. A measure of your ingenuity

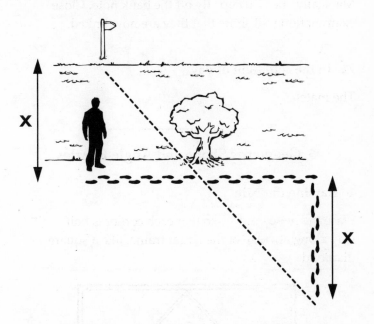

First stand directly opposite the sign post. Now walk to the tree, counting your paces. Repeat the same number of paces along the bank beyond the tree. Now turn 90 degrees and walk away from river, counting your paces. When the tree and signpost are exactly in line, stop. The number of paces retraced in a straight line back to the bank equals the distance from your original spot to the signpost.

3. Try angling

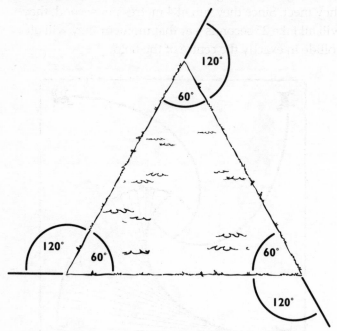

Draw a triangle and extend one of the lines beyond the angle. You have produced an exterior angle to the triangle. Exterior and interior angles on a straight line must equal 180 degrees. The triangle has three angles, which total of 180 × 3 = 540 degrees. Take the interior angles away: 540 – 180 = 360 degrees. In walking around the lake, you walked around the three exterior angles.

4. Square dogs

Each dog's target will always be running at right angles to its chaser. Because they are all moving, they will all run in the same curve. As they run at right angles to their target, each dog will run exactly 100 metres before they meet. Since they run at 4 metres per second, they will all take 25 seconds – at that moment they will all collide in exactly the centre of the field.

5. Cutting corners

No, it is not possible. Your dominoes will each cover two squares, a black and a white square. A chess board has one long diagonal of black and one long diagonal of white squares. The missing corner squares, on the same diagonal, are the same colour.

30 dominoes will cover 60 squares all in pairs of black and white. You will then have 2 squares the same colour, which cannot be covered with the remaining domino.

6. House hunting

Approaching a square house, you will be able to see
one side or two sides. So what are the odds regarding
whether you can see one or two sides from any given
position?

Imagine the house as the central square of a noughts and
crosses board. The house will be surrounded by eight
squares. From the central side squares you will see one
side of the house. From the corner squares you will be
able to see two sides of the house. So the odds of seeing
one or two sides are absolutely even.

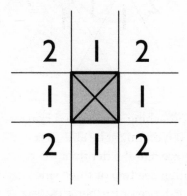

What if you are approaching a regular three-sided
house? What are the odds that you can see one side as
opposed to two sides?

This time draw or imagine a regular triangle surrounded
by other triangles. Count the surrounding triangles and
you will find there are twelve. From nine of the twelve
you can see only one side of the triangular house. From

just three you can see two sides of the triangular house. So the odds of seeing one side as opposed to two are 3 to 1.

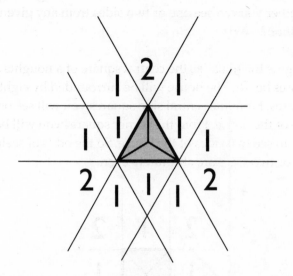

If you are approaching a five-sided house, if you are very close to it, you might be able to see just one side. But once you are half the depth of the house away from it, you will either see two or three sides. So assuming you are that distance or further away, what are the odds of seeing two or three sides?

A simple way to find the answer is to imagine another person at a point exactly on the other side of the house and the same distance away. Now if they can see two sides, you can see three, and vice versa. So providing you are more than half the depth of a five-sided house away, the odds of seeing either two or three sides are absolutely even.

7. Along the same lines

Imagine one side fixed and the other attached to one end and swinging around. When the two lines are at right angles, the subsequent triangle will have the largest possible area. The length of the third side will be the square root of 2 units long.

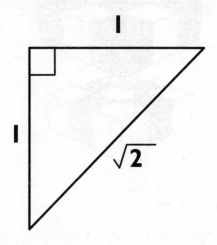

8. Strictly for cubes

No, you could not. Look at the original cube and count the black and white cubes. In each layer there are 5, then 4, then 5 white cubes – a total of 14. In each layer there are 4, then 5, then 4 black cubes – a total of 13. So if you make 13 double cubes, you will be missing a single white cube. But the centre cube has to be black.

9. Strictly for squares

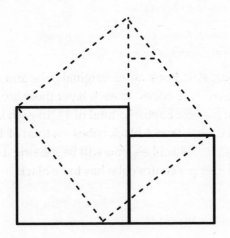

It is amazing that this solution works for any two squares of any size. You need to construct two lines. Each starts at the top outside corner of the two squares and they meet at a right angle somewhere along the bottom edge of the larger square. The question is, where along that line?

From the bottom left corner of the larger square, measure a distance equal to the side of the smaller square along the base line and mark a point.

From that point draw two lines to the top outside corners of the two squares. The lines will form a right angle where they meet. Cut along those lines.

Now you can simply rearrange the five pieces to form one complete square.

10. Just passing through

The line passes through six single squares.
It passes through all six 2 × 2 squares.
And two 3 × 3 squares.

So it goes through 14 squares altogether.

11. Getting the drop on the count

The answer is Pi (π), the circumference of a circle divided by the diameter of a circle, or the distance around the circle divided by the distance across it.

Explaining why is quite complex, but here goes:

Remember that the lines are twice the length of a pin
    apart.
Imagine if every dropped pin landed exactly vertical
    and parallel to the lines? Then the only pin touching a
    line would have landed directly along a line.
Imagine if all the pins landed horizontal to the lines.
    Now the chances would be $^{50}/_{50}$ that a horizontal pin
    crossed or did not cross a line.
Now imagine we revolved each fallen pin around its
    head, to sweep out a circle.
If the head of the pin was exactly half way between
    lines, then the circle formed would never cross a line.
If the head was on a line, then every position in the
    circle would touch the line.

So, from this Buffon suggested that, by the law of
averages, the number of pins touching a line multiplied
by π ($22 \div 7$) should give a figure very close to the total
number of pins dropped. He proved this to be so, but he
dropped many thousands of pins in doing it.

12. Look out! There's a bear behind!

The bear was white!

To walk 5 miles south, 5 miles east and 5 miles north, the
man would usually complete 3 sides of a square and be
5 miles east of where he started. So how could he have
arrived at his starting point? He must have started from
the North Pole.

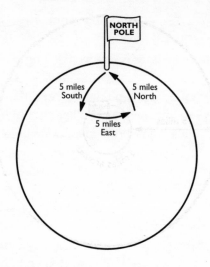

First he walked 5 miles south and then turned left. If he now walked 5 miles east, he would walk along a line of latitude. During this walk, he would always be 5 miles south of the North Pole. So, by then running 5 miles north, he must arrive back at the North Pole.

13. Same question from the other end

No, you cannot explain his starting point exactly, but it is possible to take an identical journey and see a penguin. His starting point would need to be close to the South Pole.

Let's assume we choose a circular line of latitude very close to the South Pole, which measures exactly 5 miles around it. Now if the photographer's camp was any point exactly 5 miles north of this line, he could walk 5 miles south, turn and walk 5 miles east (or west, for that matter) around the South Pole to the same point and then walk 5 miles north to his camp.

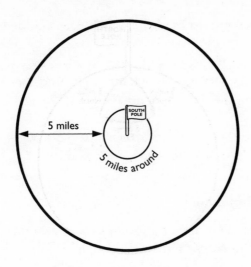

But he could also do the same thing if the line of latitude was nearer the South Pole, so that the distance around it was, say, 2.5 miles. In that case he could walk 5 miles south, then twice around the latitude ring before going 5 miles north to his camp.

So, there are an infinite number of starting points, all close to the South Pole, where the 5 miles south, east, north trip would bring him back to the same starting point.

14. By the light of the silvery lune

Normally for Pythagoras' theorem, the square on the hypotenuse is drawn outside the triangle.

But in this case the large semicircle was drawn on the other side of the hypotenuse and the three semicircles overlap to create the lunes.

Knowing Pythagoras' theorem, what can you deduce?

You can see that each white area occurs in two semicircles. The area of each lune must therefore equal the area of the shaded triangle below it.

These lunes were the first ever shapes with curved edges whose areas could be stated with certainty.

15. A elliptical orbit

The method for drawing an ellipse of any desired dimensions is easy, but seldom taught.

Here's how to draw an ellipse 12 cm wide by 8 cm deep …

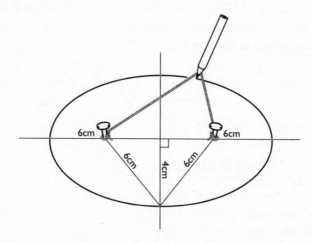

Draw a horizontal line and a vertical line crossing it. Measure 6 cm each way from the centre to find your 12 cm line. Measure 4 cm up and down from the centre to find your 8 cm line. Now, where along the 12 cm line do your pins go? It doesn't matter right now.

The string drawing your ellipse must stretch from one pin along the horizontal 12 cm line to one end and then back to the other pin. How long is that string? Well, it reaches between the two pins and passes from one pin to the end of the line twice.

So, the string needs to be exactly 12 cm long to draw an ellipse 12 cm wide. So, tie a thin string around a pin and, exactly 12 cm away, tie the string to the second pin. But where along the line will the two pins go to get an ellipse 8 cm deep?

Stretch the 12 cm string so that the middle reaches the bottom end of the 8 cm line and the two pins are still on the 12 cm line, equidistant from the crossing line. Or, the pins each need to be on the 12 cm line, exactly 6 cm from the end of the 8 cm line. Now you have the points where the pins must go to draw the ellipse you planned.

16. Upsetting Lewis Carroll's applecart

With the four wheels in identical positions, the cart would lurch up and down. With the front wheels on their ends and the rear wheels on their sides, the cart would still lurch forwards and backwards but not nearly as much. With the wheels on their ends on one side and on their sides on the other, it would rock from side to side. By trial and error, you will find that the cart runs most smoothly when the wheels are set a quarter of a turn apart, with diagonally opposite wheels half a turn apart.

17. Curves of constant width

This is how to make an enclosed curve with a width that is always the same ...

First draw any four lines, each of which crosses the other three. You could photocopy the design included here.

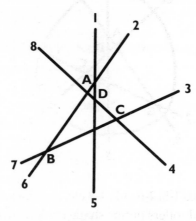

Now choose any crossing point A, where two lines form an angle. With the point of the compasses at A, draw a curve joining the two lines that cross at A. Number the two ends of the curve 1 and 2.

Now trace the 2 back to where it forms an angle with the next line to the right of it. Call the crossing point B. Next, with B as centre and B2 as radius, draw a curve to the next line and number that point 3.

Now trace line 3 back to where it makes an angle with the next point. Call that point C and with C3 as radius draw an arc to the next line, calling that point 4. Then find point D and draw arc 4-5. Now go back to point A and draw arc 5-6. Go to point B and draw arc 6-7. Go to point C and draw arc 7-8. Lastly, with D as centre point, draw the arc 8-1. Your shape has a constant width.

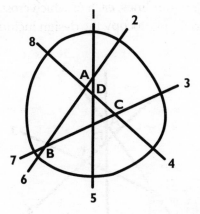

As proof, lines 1-5, 2-6, 3-7, 4-8 etc. are all exactly the same. Several rollers of this shape could be used to move a plank and it would move smoothly forward, with no up and down movement, as though it were on round wheels.

18. Of grave concern

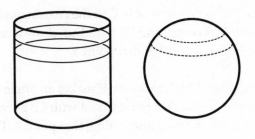

The area of the cylinder wall and of the surface of the sphere are exactly the same. It sounds amazing, but it is true. This is very useful fact in making maps or measuring areas on a sphere.

The area of a sphere's surface between two parallel lines is difficult to measure because the sphere's surface curves. But it is exactly the same area as the area of the cylinder wall, between the same two horizontal lines.

To measure a round, pole cap of a sphere would seem to be difficult. But it is easy to calculate the area of the band of a cylinder wall above that line, and the two areas, band and cap, are exactly the same.

19. A new train of thought

Certain parts of a train *are* going backwards while a train is going forwards. But the parts are changing all the time. We'll explain by looking at the wheels.

All train wheels have a flange which keeps them on the track. But as the train moves forward the part of the wheel touching the line at any given moment is stationary. A single point on a wheel's edge running along a line draws out a path called a cycloid.

However, the path of a single point on the edge of the flange of the wheel describes a path with loops in it. To make the shape of the loop, the point needs to be going backwards when it is below the level of the line. So those points on each wheel below the level of the line are going backwards, even when the train is going forward at top speed.

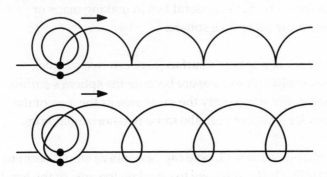

20. The quickest way down

Galileo, and later the Bernoulli brothers, experimented with Cycloids and discovered some quite surprising facts.

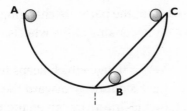

At some time in their maths schooling most people are taught that the shortest distance between two points is a straight line – correct. But that knowledge can lead to getting the wrong answer to this question.

The ball that takes the longest time to get to point X will always be C. Although the route is a straight line, the friction of ball on slope slows it down.

So, between balls A and B, surely it will be B that gets to X the quickest?

Well, in truth both A and B will arrive at X at exactly the same time.

It seems impossible, or at least improbable, but it is true. Ball A will encounter no friction on release as it moves directly downwards. The friction will then gradually increase. Ball B will start with quite a lot of friction, meaning that no matter how close it is to point X at the start, it will still take the same time as ball A to reach it .

In fact, a ball will take exactly the same time to reach X from any starting point on the cycloid curve. Also, when the two balls meet, the effect is perhaps surprising.

Both balls bounce, but the ball that has travelled further bounces less than the ball that has made the shorter trip. The two balls go up the slope away from X and then, as gravity stops them, they return – only this time as they collide at X, the slower bounces more and the faster bounces less. This continues for a couple of bounces until impetus is lost due to friction and the two balls come to rest. It is the same effect as you see with Newton's Cradle.

21. Target practice

Both shaded areas are exactly the same. The inner shaded circle has a radius of 3 units. $(3^2 = 9) \times \pi$ gives you the area. The outer ring is $(5^2 = 25) \times \pi$, less $(4^2 = 16) \times \pi$, which is also $9 \times \pi$.

22. The Mesolabe Compass

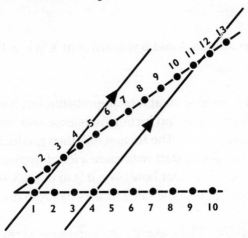

Let's imagine that you wish to multiply 3 by 4? I know that is easy, but this is just a demonstration. To carry out this operation, from where the two lines meet, simply count 3 units up the angled line and 1 unit along the base line. Draw a line through the two points, and you have a picture of $1 \times 3 = 3$. Wouldn't it be wonderful if all maths could be done in pictures?

So, to find 4×3, draw a new line passing through the 4 point on the base line. Draw the line parallel to the 1-3 line. Amazingly, it will pass through the 12 point on the upper line, showing that $4 \times 3 = 12$.

Lines drawn parallel to the 1-3 line through each base point will give the powers of 3. $1 \times 3 = 3$, $2 \times 3 = 6$, $3 \times 3 = 9$ etc. Using decimals you can show that $3 \times 3.5 = 10.5$ and so on. Also, all lines are in perfect proportion. So the $4 \times 3 = 12$ line is 4 times longer than the $1 \times 3 = 3$ line.

Descartes showed that you can have different number progressions on the upper line. Let's imagine, for example, that the upper line has units of 7, 14, 21, 28, etc. The same procedure gives successive multiples of 7 and even decimals of 7.

Division is equally simple. Take our example in reverse – to divide 12 by 4, draw a line passing through 12 on the upper line and 4 on the base line. Now draw a parallel line through the 1 on the base line. It will pass through the 3 on the upper line, proving that 12 ÷ 4 = 3.

23. Getting to the root of the problem

Here's how you find the square root of 9 …

Along your base line, count along to the 9 point. Now, here is the magic bit – add 1. Why? Because that is what makes it all work! Now you are at point number 10. Next find the halfway mark, or point 5.

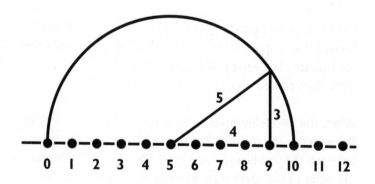

Now, with a pair of compasses, taking point 5 as centre and with a radius of 5 units, draw a circle. The curved arc will pass through the 0 and the 10 points on the line. Now draw a perpendicular (straight up) line from the 9 to where it meets the curved arc. The length of this upright line will be exactly the square root of 9, or 3 units.

The same procedure will give the square root of *any number*. Descartes chose to place this idea and that of the Mesolabe Compass in the introduction to his book *The Geometry*, and the reason he did so is obvious – together they illustrate perfectly the wonderful power of simple maths.

Here is a true story. In around 1880 a wealthy fool bet £500 that the earth was flat. Alfred Russel Wallace, who shared the discovery of evolution due to natural selection with Charles Darwin, used the method we have just discussed to prove the Earth was indeed round and even to measure the diameter of the Earth.

On a calm day, he took two punts to the Old Bedford level, a drainage canal in East Anglia. He had a telescope set 1 metre above the water on each punt. The punts were then pulled apart.

When the two telescopes were in line with the surface of the water in the middle, he simply halved the distance between the punts and squared it. His evaluation of the diameter of the earth was accurate to well within 1%.

24. Divide and rule

Can you divide any line into 7 equal parts (or indeed any number of equal parts) without measuring the line? Yes you can.

Draw a new line forming an angle with the first line.

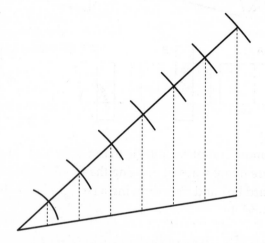

Now, with a pair of compasses, select a radius distance which you guess to be around $\frac{1}{7}$ of the length of the first line and, starting from the point where the lines meet, mark off 7 new points along the angled line. With a ruler, join the 7th mark to the far end of the original line. Now, draw parallel lines through the other 6 marks along the new line, each passing through the original line. These parallel lines will divide the original line into 7 parts of identical length. As with most geometric tricks, it is so easy once you know how.

25. The spider and the fly

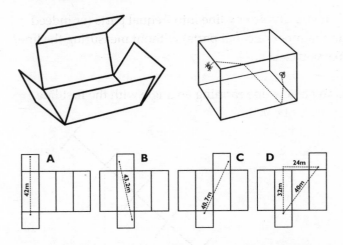

This is more complex a puzzle than you would think. There are many ways of opening the room out like a cardboard box and then drawing a straight line between spider and fly.

Look at A first – here it is easy to calculate that the spider travels 42 meters. For B, C and D, you can calculate the distance using Pythagoras' theorem – for each, draw a triangle with the path as the hypotenuse …

For B, the sides are 42 and 10 with squares 1764 + 100 = 1864 with a sq root of 43.17 m.

For C the sides are 37 and 17 – 1369 + 289 = 1658 sq root 40.72 m.

For D the sides are 32 and 24 – 1024 + 576 = 1600 sq root exactly 40 m.

So although it doesn't look like it, route D would be the quickest route. I bet the fly will have flown off long ago.

6. Dear Old Faves from Days Gone By

1. The sweet truth

Open the mixed jar and take a sweet. As all the labels are wrong it cannot be mixed, so you have located one sweet. Say you pull out a jelly baby – now the wine gum jar cannot contain wine gums or jelly babies. So it must contain mixed sweets. The jelly baby jar must contain wine gums.

2. A later edition

If page 19 is missing, page 20 will be on its reverse side and also be missing. But if there are 28 pages in total, there would be 8 pages after 20 and 8 pages before the other missing pages, so pages 9 and 10 will also be missing.

3. Half full, half empty

Simply empty the contents of the second full glass into the already empty fifth glass. Replace the second glass and the glasses are now alternately full and empty.

4. Beauty is only skin deep

30 minutes is the fastest time possible. One of the beauticians does a face pack and three manicures, the other does two face packs.

5. A leap of faith

After 7 days, the frog has climbed 7 metres. That leaves him 3 metres from the top and in one more jump on the 8th day, he is out.

6. Don't get this one rung

After 4 hours, the boat will be floating 120 cm higher, but as the boat is floating, the same 12 rungs will still be in view.

7. Combined assets

Alf stood in front of the bank, while Bill went to stand behind the bank. Now they had all the money in the bank, between them. It should be said that these days not many banks will have as much as a million pounds in cash in them.

8. The last word in puzzles

I like this puzzle because finding the answer takes you almost right through the alphabet.

R, S, T, and U give 'rust' or 'ruts'. This is the only consecutive four letter group in the alphabet from which it is possible to form a word in this way.

9. Is there a short cut?

As the centre cube has 6 faces, it must require 6 cuts. No amount of stacking the pieces between cuts will reduce the required 6 cuts.

10. A rather nasty turn

You simply say to either of them, 'If I were to ask the other chap which is the way to Heaven, what would he say?'

The angel must admit the other would lie. The Devil's man would lie in his answer.

So, whatever answer they give, the road to Heaven is the other way.

11. An up and down kind of existence

It is an absolute certainty. The monk must be at some point on the road at exactly the same time on both days. Imagine two monks going each way on the same day – they must pass, somewhere, and be at exactly the same point on the road at that moment.

12. Bending the rulers

No you cannot. It is, in two words, im-possible.

Imagine that rather than fold the paper ruler, you slide it along the school ruler so that the start of the paper ruler, is above the 10 cm point on the school ruler. 10 cm of the paper ruler is now hanging beyond the end of the school ruler.

Next, fold it back so the fold lies on the end of the school ruler. Now the 25 cm mark will be in line vertically on both rulers. No matter how complex your folds are, as long as the paper ruler is completely over the school ruler, two identical points on both rulers must always be in line.

13. Losing one's bearings

Once again, it is a certainty. The crumpled map *must* have a point over an identical point on the flat map. It is just like the paper ruler puzzle but in three dimensions.

What if the two maps are of a different scale? Once again there must be a point in line vertically on both maps, provided the crumpled map is on, and not overhanging, the flat map.

Just get a map of London, roll it into a ball and throw it on the ground anywhere in London. A point on the map must be directly over that actual point in London.

14. Short changed

There is no missing £1. The men paid £27 – £25 for the meal and the £2 the waitress took as a tip. The error is in then adding the £2 to the £27 when it is already part of the £27.

15. Talking telephone numbers

With any three-digit number, if you reverse it and take the smaller from the larger, the answer is always a three-digit number, with a 9 in the middle and the end two digits adding to 9. When you reverse that and add the two together the answer is always 1089.

If your three-digit number has end digits just one apart, when you reverse it and take the smaller from the larger, the answer is always 99.

For the trick to work you must treat it as a three-digit number, calling it 099. Adding 099 to its reverse, 990, still gives 1089.

16. The leaping loop

As you close your fist, guide all four fingers through the band with the other hand. Now straighten out the fingers – the band jumps from one pair of fingers to the other.

In the second version, the band around the ends of the fingers makes no difference. Push all four laced fingers through the band as you clench the fist. Straighten the fingers and the band has now made the seemingly impossible leap.

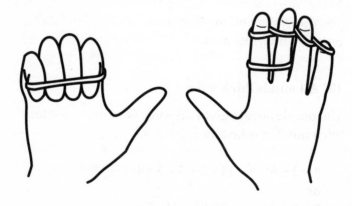

17. The mind reading rip-off

You rip the paper into three pieces one way. You then stack the three pieces together and rip them in three the other way. You now have nine pieces of paper. However, just one – the centre piece – will have four torn edges.

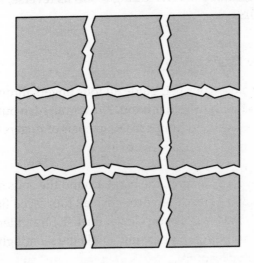

Make sure you know which person gets that piece. Now you can easily find the piece again and identify the person who wrote on it.

18. All minds think alike

The puzzle always gives the same answer, no matter what number is first thought of, so:

$3 \times 2 = 6 + 8 = 14 \div 2 = 7 - 3 = 4$

or

$7 \times 2 = 14 + 8 = 22 \div 2 = 11 - 7 = 4$

So they are forced to find a country beginning with the letter D. They could choose the Dominican Republic, but almost everyone chooses Denmark. The second letter is E, and almost everyone thinks of an elephant – not an emu. In Africa elephants often roll in red dust, but of course generally everyone thinks of elephants as grey.

19. Where on Earth?

Anywhere, for almost half the year.

From midwinter and the shortest day to midsummer and the longest day, as the days grow longer, so the sun rises slightly earlier and therefore rises twice, just, in each 24 hours.

20. This might give you a furrowed brow

None – the plough being pulled behind the ox erases all the tracks. This trick question was actually found on an ancient Babylonian clay tablet.

---

7. Even More Thoughtful Thinking Puzzles

1. Jobs for the boys

You can work out who has which jobs like so …

From 6, the painter must be Alf.
Alf cannot be the shelf stacker (3) the driver (4), or the
   musician or gardener (2).
So Alf must be the painter and the hairdresser.

Charlie must be the gardener (5)
Bill must be the musician (2) but not the driver (1).
So Bill must be the musician and the shelf stacker.
And Charlie must be the driver and the gardener.

2. If truth be told

Only one tribe always tells the truth.

It cannot be A's tribe, as he wouldn't say someone else 'always tells the truth'.
B also can't be a total truth-teller, for the same reason.
So you know C is the total truth-teller, without him needing to say a word.
But A said 'C always tells the truth', so A must have told the truth this time.
That means that A is an Eachwaybet, and so, B must always lie.
And saying 'A always tells the truth' is a lie, which proves it.

3. Angels with dirty faces

As they get up, relieved that they are OK, the boys look at each other. The clean-faced lad sees his pal's muddy face and thinks, 'My face must be dirty too.' The muddy lad, on the other hand, looks at the clean-faced lad and seeing nothing wrong. So, he goes straight to class while the clean-faced lad decides he needs a wash.

4. This could drive you dotty

Yes. It must always be possible.

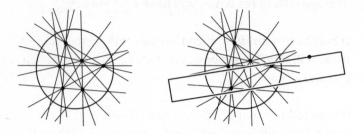

The method is simple to explain, though it would be difficult to do with 500 dots. You need to draw lines which pass through every possible pair of dots. The lines all need to continue beyond the dots outside the circle.

No matter how many lines you draw, there will be spaces between the lines. So choose any space outside the circle and place a dot in the space. Now, with the edge passing always through that dot, swing the ruler round and count every dot the ruler passes inside the circle.

As your dot is not on a line, your ruler must pass through each dot inside the circle individually. Keep going until the ruler has passed through exactly half the dots. Now you can draw a new line that will divide the dots in the circle exactly in half.

5. A king's kith and kin

Each wife might have either a boy or girl first. So on average 50% of his wives will have a single boy.

If half have a girl first, then the odds will be that half of them will then have a boy and stop. So 25% will have a girl and a boy.

It then follows that 12.5% will have 2 girls and 1 boy, then 6.25% will have 3 girls and 1 boy and 3.125% will have 4 girls and 1 boy.

So it works out that the law of averages will win in the end, and the chances of having 50% boys and 50% girls will be maintained – on average that is.

6. Happy birthday to you two

The answer (almost as well known as the question) is 23. But why?

If there are 2 people the odds of them both having the same birthday is $\frac{1}{364}$. If there are 3 people then there are three pairs of birthdays and the odds are $\frac{3}{362}$. For 4 people there are 6 pairs so the odds are $\frac{6}{359}$. For every extra person, the odds of at least 2 matching birthdays gets smaller.

I have a friend called Frank Duckworth who shared grotty student digs in Liverpool with me 50 years ago. He has spent his life studying statistics, and devised with his mate, Tony Lewis, the Lewis-Duckworth cricket scoring system for deciding cricket matches half-finished and abandoned because of rain.

Frank explained that with the 'same birthday' question, there is a simple formula which gets you to the answer 23. You multiply the days in the year by 1.4 and then find the square root:

So 365 × 1.4 = 511, the square root of which is 22.60.

So with 23 people in a room, the odds are slightly better than even that 2 will have the same birthday.

7. A hair-raising question

For our hairs on a head problem, if we assume the average number of hairs is 80,000, then:

80,000 × 1.4 = 112,000, the square root of which is 334.66.

So, according to my statistically minded pals, Frank Duckworth and Frank Haigh of the Statistical Society, once you have 335 people in a crowd, then the likelihood of two having the exact same number of hairs on their head, is just slightly better than 50⁄50.

If you feel the need to put this to the test then please do start head hunting and counting, but do be careful – if you loosen a single hair during the count, you'll have to start all over again. Then again, do you count a split end as one or two, or am I splitting hairs?

8. The answer is buried in the sand

As Zoe could bury me in 10 minutes, she could bury
6 dads in 1 hour. Nick would take 15 minutes and could
bury 4 dads in 1 hour. Dan taking 30 minutes could only
bury 2 dads in 1 hour. So, altogether the 3 of them could
bury 12 dads in 1 hour. Which means that to bury 1 dad,
yours truly, they would take $60 \div 12 = 5$ minutes.

9. He hadn't got a sausage?

The 3 sausage roll man has no right to claim half the
money. But the 5 sausage roll man asking for £5 was
demanding less than he deserved. The wise old man
explained, thusly:

To divide the 8 sausage rolls between three men, they
    had to cut each one into 3 pieces.
That made 24 pieces, so each of the men ate 8 pieces.
That meant the 5 sausage roll man had contributed
    15 pieces and eaten 8.
While the 3 sausage roll man had started with 9 pieces
    and had eaten 8 himself.
So the 5 sausage roll man, for the 7 pieces he had
    contributed to the business man, deserved £7.
The 3 sausage roll man contributed only 1 piece and was
    only entitled to £1.

10. Weigh above your head?

The solution is quite complex …

Number the coins 1 to 12 and divide them into three groups of 4.

First weigh 1, 2, 3 and 4 against 5, 6, 7 and 8.

If they are even, then the dud is 9, 10, 11 or 12.

Secondly, weigh 9 and 10 against 11 and any other.

If they are even, the dud is 12. Weighed against any other shows whether it is lighter or heavier.

If there is an imbalance then weigh 9 against 10.

If they balance the dud is 11. The second weigh will have told you whether it is lighter or heavier.

If the balance dips the opposite way to the second weigh, the dud is 10.

If the balance is the same as weigh 2, the dud is 9.

Now let us assume the first weigh 1, 2, 3, 4 against 5, 6, 7, 8 was not even.

Note which way the pan dipped.

For the second weigh, weigh 1, 2, 5, 6 against 3 and any three from 9 to 12.

If they now balance, the dud is 4, 7 or 8.

Weigh 7 against 8.

If they are even, the dud is 4 and the first weigh tells you whether it is lighter or heavier.

The same dip as weigh 1 tells you it is 8.

The opposite dip tells you it is 7.

If weigh 2 shows the same imbalance, the dud is 1 or 2.

Weigh 1 against 2. The same dip and the dud is 1, the opposite dip and the dud is 2.

If weigh 2 shows an opposite imbalance, the dud is 5, 6, or 3.

For weigh 3, weigh 5 against 6.

If they are even the dud is 3 and the first weigh tells you whether it is lighter or heavier.

If the dip is the same as weigh 2, it is 5.

If the dip is opposite to weigh 2, it is 6.

Phew!

11. A monkey puzzle

The sack also rises and they stay at the same level.
The monkey will only rise from the ground by half the
length of rope he climbs. Also, if he climbs down the
rope, the sack will descend as well.

12. Blind faith

Each reasons as follows ...

They have been told that all three cards cannot be black.

Now, can two cards be black?

No, since a person seeing two black cards would know
theirs was white.

So can one be black?

No, as someone seeing a black and white card would
have no idea of the colour of their own card, and they
were told that they would be able to identify the colour
of their card.

So all the cards had to be white.

13. Hey diddle diddle, there's a hole in the middle

The question does not state the volume or diameter of the ball or of the hole. So the only measurement mentioned must be all you need to work out the answer.

If you don't know the holes diameter, any diameter hole must give the same answer. In that case, let's assume the hole has no width or diameter at all.

So the volume of the ring would equal the volume of a ball of 6 cm diameter (and 3 cm radius).

As the volume of a ball is ⅓ πr^3 which gives us:

$$\text{⅓} \times 22 \div 7 \times 3 \times 3 \times 3 = 113\text{½ cm}^3.$$

Let us assume the ball is the size of the earth, with a hole through the middle, leaving the wall of the hole just 6 cm high. This huge very slim ring would be 12,700 kilometres in diameter.

But it would be so thin it would still only have a volume of 113½ cm³.

14. An amazing chain of events

To separate all 7 links, she would need to cut alternate links – 2, 4 and 6.

However, if the inn keeper is happy to exchange links then she needs only have one link cut, the 3rd one from one end, giving a single, a double and a 4-link chain …

Now she can pay the single link on day 1.

On day 2 she takes the single link back and gives the 2-link piece.

She gives the single link on day 3.

On day 4 she takes all 3 back and gives the 4-link piece.

On day 5 she gives the single link again.

On day 6 she takes back the single link, and gives the double link.

On day 7 she gives the single link by which time, hopefully, she will have been rescued.

15. Boys will be boys and vice versa

Well they can't both be telling the truth as 'at least one of them is lying'. If the boy is lying, then he is a girl, but at the beginning we said a boy and a girl were sitting on the steps. So if the boy is lying, so is the girl and vice versa.

So they must both be lying, which means the redhead is the boy and the dark-haired person is the girl. Who could they be? This smacks of the best man and bridesmaid at the royal wedding, which had just happened when I wrote this.

16. How card sharp are you?

No, it is not fair. Once you turn up a card if, say, it's black then of the other 3 cards only 1 is black and the other 2 are red. So in choosing 2 to be the same, for every 1 chance of success you have, I have 2. Your chance of winning is always only $\frac{1}{3}$.

17. Well worth the weight

Bill weighs 48 kg. Could he be the lad who weights 8 kg less than the heaviest?

Well, that would mean the heaviest weighing 56 kg.

Then from 160 kg, take 48 kg and 56 kg that leaves another 56 kg boy.

So, Bill at 48 kg, is not 8 kg less than the heaviest. So who is?

Taking Bill's 48 kg from 160 kg leaves 112 kg.

Splitting 112 kg with a difference of 8 kg gives 60 kg for the heaviest, and 52 kg for the other. Both are heaver than Bill.

Alf weighs less than boy in school shoes, so he weighs 52 kg.

The heaviest must be Charlie at 60 kg and he is in school shoes.

Charlie weighs more than the lad in trainers, who must be Bill.

Alf, at 52 kg, must be wearing sandals. Sorted!

18. Ill-gotten gains

This looks like an algebra problem, where you compare equations, but don't 'e worry – it be easier than that.

Let's take the second bit of info first:

We know that ½ Sid's share + ½ Pete's share + 1½ Jack's share = 1000.

So ½ Sid's + ½ Pete's = ½ Jack's share.

So, all of Sid's share and all of Pete's share would equal all of Jack's share.

So Jack's share must be 500 pieces of eight.

From the first statement, ½ Jack's share + ½ Sid's share +
   3 times Pete's share = 1000.
So, Jack's 500 + Sid's share plus 6 times Pete's share =
   2000.
Sid's share plus 6 times Pete's share = 1500.
So, Sid's share is 300 and Pete's is 200.

19. Age concern

This is easier than it sounds.

'My son is as many weeks as my grandson is days.'
So the son must be 7 times older than the grandson.
Then he said, 'My grandson is as many months as I am
   years.'
So the man is 12 times as old as his grandson.
So, assuming the grandson is 1, the son would be 7 and
   the dad 12. Added together 1 + 7 + 12 = 20.

But their combined age is 100, not 20.

So multiply everything by 5 and you have: the grandson
is 5, the son is 35 and the dad is 60.

$$5 + 35 + 60 = 100.$$

20. Crazy golf

With the 125 yard driver and 75 yard wedge, his score
was:

   300 – 3 drivers and a wedge back, 250 – 2 drivers,
   200 – a driver and a wedge, 325 – 2 drivers and a

wedge, 275 – a driver and 2 wedges, 350 – a driver and 3 wedges, 225 – 3 wedges, 375 – 3 drivers, and 400 – 2 drivers and 2 wedges. That's a total of 28 shots.

With the 200 yard driver and the 25 yard wedge, his score was:

300 – a driver and 4 wedges, 250 – a driver and 2 wedges, 200 – a driver, 325 – 2 drivers and 3 wedges back, 275 – a driver and 3 wedges, 350 – 2 drivers and 2 wedges back, 225 – a driver and a wedge, 375 – 2 drivers and a wedge back, 400 – 2 drivers. That's a total of 29 shots.

But his best two clubs of all were a 125 driver and a 100 yard wedge, which scored:

300 – 3 wedges, 250 – 2 drivers, 200 – 2 wedges, 325 – a driver and 2 wedges, 275 – 3 drivers and a wedge back, 350 – 2 drivers and a wedge, 225 – a driver and a wedge, 375 – 3 drivers, 400 – 4 wedges. That's just 26 shots in total.

What a player! Luke Donald, eat your heart out.

21. A cruising conundrum

You will pass another ship every half day, and in 5 days you will pass 9. But on leaving Southampton, another will be arriving, and on arriving in New York, another will be leaving. So you will see 11 other ships doing that trip and the service will require 12 ships.

22. An age-old problem

Here is the question again. Alf is twice as old as Bert was when Alf was as old as Bert is now. When Bert is as old as Alf is now, their combined age will be 63. How old are Alf and Bert now?

Start at the end with the 63. This divides by 7 and 21. The 7 is the secret.

'Alf is twice as old as Bert was when Alf was as old as Bert is now.'

So at one time Alf was twice Bert's age, or $\frac{2}{1}$, and combined they were 3 units in age.

When Bert is or was 2 units, Alf is or was 3.

When Bert is 3 units, Alf will be 4 units old. $3 + 4$ units = 7 units. $63 \div 7 = 9$.

So, when their combined age is 63 Alf will be $4 \times 9 = 36$, Bert will be $3 \times 9 = 27$.

23. One more river to cross

There are 4 tourists and 4 vampires on opposite banks of the river. The boat holds 3 people, so it cannot start on the tourists' side, as 3 tourists would be outnumbered by 4 vampires when they crossed. So, the canoe must start on the vampires' side. Then the solution goes like this:

The vampire rower brings 2 mates across – VR, V, V.

Now, leaving 2 of each behind, both rowers cross with a tourist on board – TR, VR, T.

Now the tourist rower brings a non-rowing vampire
   back – TR, V.
Then the 3 remaining tourists cross – TR, T, T.
And, lastly, the vampire rower – VR – crosses alone, on
   his own and by himself.

Just 5 crosses and no blood spilt, although the stakes
were pretty high.

24. The three colour map theorem

This question needs no explanation – it *is* possible, and
there are many ways of doing it. Did you find one?

25. Everyone to the boats

There are in fact millions of solutions, as once you have
a schedule, you can simply change any 2 or 3 girls over
and you have a new version.

But to not include an answer, which was my first
intention, would have been very sneaky. So here is just
one solution …

Sun.	Mon.	Tues.	Wed.	Thurs.	Fri.	Sat.
01, 06, 11	01, 02, 05	02, 03, 06	05, 06, 09	03, 05, 11	05, 07, 13	11, 13, 04
02, 07, 12	03, 04, 07	04, 05, 08	07, 08, 11	04, 06, 12	06, 08, 14	12, 14, 05
03, 08, 13	08, 09, 12	09, 10, 13	12, 13, 01	07, 09, 15	09, 11, 02	15, 02, 08
04, 09, 14	10, 11, 14	11, 12, 15	14, 15, 03	08, 10, 01	10, 12, 03	01, 03, 09
05, 10, 15	13, 15, 06	14, 01, 07	02, 04, 10	13, 14, 02	15, 01, 04	06, 07, 10

26. Josephus is not one of us

The magic number is 9. Try it.

Remember, when a letter has been eliminated, it is never counted again as you go around. You might photocopy the page and try it, crossing out those eliminated as you go.

The 15th B to go is the middle B at the 3 o'clock position. The 1st A to go is the 2nd A at the 10 o'clock position.

Now, does it work according to the Japanese tradition? Try it. Stop at the 14th B, the first B you come to at around the 8 o'clock position. Now count backwards and eliminate every 9th person. The As disappear one by one, until the last letter to remain is the single B at 3 o'clock. So, the Japanese version does work with this particular circle of 30.

---

8. Party Puzzles and Tricks to Show Off With

1. Get your money out

To do this you simply have to blow hard down the inside of the glass. The blast of air makes the large coin turn over. The small coin shoots up and out of the glass and the large coin settles back where it was.

BLOW

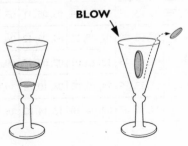

2. The bend in the fork

As is often the case, the important part of the trick is done hidden behind the hands.

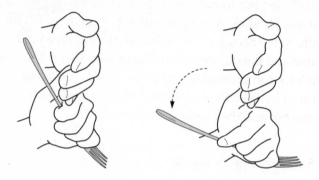

You need to grip the fork with one thumb behind the fork handle. As you push down, allow the thumb to relax and open backwards. This lets the handle of the fork bend backwards naturally as you press down. As you press down, give a grunt of effort, for effect. It will appear that you have bent the fork, when you haven't harmed it at all.

3. Blown away

Stand the bottle quite near to the candle. Now, blow hard at the bottle and, surprisingly, the candle will go out. The reason this happens is that moving air has less pressure than still air. So, your moving breath is held close to the bottle by the surrounding stiller air. It therefore hugs the bottle sides and still arrives at the candle, which is blown out.

4. Candle power

In order to blow a candle out with an empty bottle, first hold the bottle to your mouth, with your thumb actually in your mouth. Now, blow hard into the bottle and quickly stick your thumb over the opening. Tilt the bottle so the neck is close to the candle. Remove your thumb and the candle will be blown out! This is because you have compressed the air in the bottle – on being released by your thumb, the air expands again and rushes out and the candle is extinguished.

5. Small change changing

Put your left index finger on B, holding it still so as not to move it. Now, with your right index finger, move C to the right a few inches. Now flick C to the left so it hits B. A will be knocked sideways. There will now be plenty of room to place C between A and B.

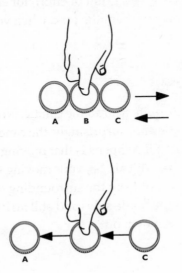

6. Chase the ace

If you have 6 cards, there are 15 possible ways of taking 2 cards. Let's assume the ordinary cards are numbered 1 to 4 and the aces are cards 5 and 6.

Any pair without an ace would have to be drawn from cards 1 to 4. There are just 6 pairs that can be made from those 4 cards, that's only 6 pairs with no aces. So of the 15 possible pairs, 8 will contain 1 ace and 1 will contain 2 aces.

So there are 9 possible pairs with an ace and only 6 pairs with no ace.

So the chances of any pair containing an ace are ½ in favour. So, you must go for an ace being one of the 2 cards you will select.

You might be interested to know what the odds are with 7 cards, of which 2 are aces. Now there are 21 possible pairs. 10 would have no ace. 10 would have 1 ace. A single pair would have 2 aces, so the odds are still in favour of finding at least 1 ace – just.

7. In black and red

Amazingly the cards always come off the top in pairs of one red card and one black card. But why and how?

The secret is, when you show the audience that the cards seem to be no longer alternately red and black, note two cards of the same colour, cut between those two cards and complete the cut. Now each pair from the top will be a red and a black. However, whether red or black comes first, no one can tell.

8. Suitably suited

This trick is almost identical to the previous one, but even more impressive. When you show the audience that after the shuffle, the cards are no longer in perfect suit order, look for two cards of the same suit next to each other. There will be at least one pair like this. Cut between them and complete the cut. Now when you take four cards from the top of the pack, they will always be one of each suit.

9. Place them any place

The answer is yes, you can. Place each domino exactly diametrically opposite to where your opponent plays. In this way, whatever space he takes up, the opposite space will be available for you, and you must win.

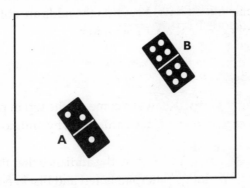

Now here is a more difficult question – if you have to play first, is there a strategy that will still ensure that you win? There is – lay the first domino in exactly the centre of the playing surface. Now wherever your opponent plays, you can echo that position in the same way as before.

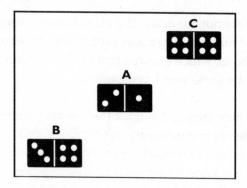

10. A good deal of dealing

The deal must have started with the player on his left, and gone around clockwise. The complete deal would then have to finish with the dealer himself. So, if he deals the cards left from the bottom of the pack, dealing to himself first, and then goes around anticlockwise, he will complete the deal and every player will have exactly the cards they would have had if he had not been interrupted.

11. Making a pile

The key here is that the opposite sides of a normal die always add to 7. So, if the pile is 3 dice tall, then all the top and bottom faces must add to 21. You simply need to subtract the number on the top of the pile from 21. If, for example, the top number on the pile is 4, just take it from 21 and you have the answer – 17.

12. You can bet it's best not to bet

No, they are not fair odds. The idea of using symbols instead of numbers 1 to 6 was probably to cover up the crafty wheeze which ensured that in the long run, those running the game won handsomely.

Let's imagine there is a single £1 bet on each number. If three different symbols come up, there are three losers, three winners and no gain for the house. With two numbers the same, £3 is paid on the double and £2 on the single – that is £5 paid out of £6 and the house keeps £1. With three dice all the same the house pays out £4, but keeps £2 profit.

For any one symbol, there are 36 possible throws in which it will be involved. 20 of those will show just one winning symbol and there is no gain for the house. Fifteen throws will be doubles and on each the house will gain £1. One throw will be a triple and the gain for the house will be £2.

Overall, taking all throws into account, the bank reaps a profit of £7.9%. That might not sound like much, but as most players will bet again with their winnings, that percentage is ample for the house to make a very handsome profit.

The true wins for a fair game should be £2 for a single, £4 for a double and £6 for a treble.

13. A toss-up between good and evil

Well, most coins have a head on one side and a more complex design on the other. Just feel each side with your thumb and you will soon be able to tell, without looking, which side your thumb is rubbing against. So, it is quite easy to toss a coin and ensure that whatever anyone calls will be wrong. Here's how ...

Listen for the call and then catch the coin in your right hand. Instantly feel the coin with your thumb to identify whether you are feeling a head or a tail. If heads has been called and you feel a head, then slap the coin down on the back of your hand and reveal it – the upper side will now be a tail.

If heads has been called and you feel a tail, it is that side you want to be uppermost. So, as you cross your hand over, simply turn the coin over in your hand. Slap the coin down tail uppermost and they will have lost again.

This is probably why in sport the coin is always tossed to land on the ground.

14. All the fraud of the fair

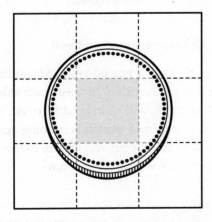

Well, this picture shows a square and a coin. The large square has been divided into 9 smaller squares. For the coin to be inside the square and not touching a line, the centre of the coin needs to be over the central square – over any other square and it must cross a line. So, the odds of a rolling coin landing in a winning position are ⁸⁄₁. Hardly seems fair? Well, fairground people have to live in the winter when you've gone, you know.

15. Every one's a winner

You can always make the number chosen lead you to the 'phantom red card' ...

If the person chooses 1 or 6, count 1 from the right or 6 from the left. Then turn the 2 of clubs to reveal that it has a red back.

If they choose 3 or 4, interpret that as the 3rd from the left or the 4th from the right. You can then turn over the downward-facing black-backed card to reveal it is the red 5 of hearts.

If they choose 2, the 2 of clubs has a red back, which you can reveal.

If they choose 5, the 3rd from the left can be turned over to reveal the 5 of hearts.

It's a bit like magic, isn't it?

16. A certain solution

Writing a digit three times is the same as multiplying the digit by 111, and $111 \div 3 = 37$.

17. Christmas past, Christmas presents

On the last day of Christmas the total number of presents received equals 12 factorial, or the total of the first 12 numbers all added together:

$1 + 2 + 3 + 4 + 5 + 6 + 7 + 8 + 9 + 10 + 11 + 12 = $
78 presents received on that day.

Over the 12 days the total number of presents received must equal the total of the first 12 triangular numbers (numbers that form a triangle if you arrange them like snooker balls). The triangular numbers in question are:

$1 + 3 + 6 + 10 + 15 + 21 + 28 + 36 + 45 + 55 + 66 + 78$
$= 364$

So that's just one short of a present for every day of the year!

18. Gone to pot

Can a snooker player make a break of 8 and pot 4 yellow balls legitimately?

Yes he can, and here's how …

The situation in the game is that there is just one red left on the table, but the last player has fouled. Now the new player is snookered on the last red, so he can claim a free ball and choose any colour to pot, which will count as a red and score 1.

So, he pots a yellow, scoring 1. The yellow is re-spotted and he pots yellow as his nominated colour, adding 2. The yellow is re-spotted. Then he pots the last red, then yellow again and is now 'on to the colours' in which case he pots the re-spotted yellow. His break is now 8 and in doing it, he has potted the yellow four times.

19. A mixture mix-up

As both glasses are still half full, as they were at the start, the amount of water in wine and wine in water must be exactly the same.

20. Card sharp

Place the ace (1) at any point and then place the 2 two points away, then the 3 also two points away, going in the same direction – and the 4 and the 5 likewise. Place the 6 between the 3 and 5 and then come back past the 5 and place 7, 8, 9 and 10 on the mid-points in order. The five sides will now all add to 14.

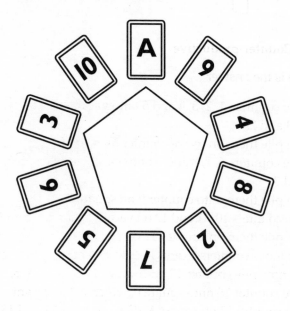

21. Party pooper

At first the blower will fail. This is because the air is not compressed by the cone, but fans out in several directions. However, it can be done. Line up the candle so that it is on the extended line of the cone's lower side. Now the blower will have no difficulty blowing out the candle. Snuff said.

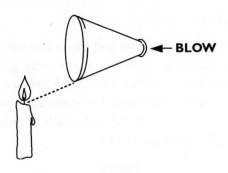

← **BLOW**

22. Counter-productive

Here is the answer …

Move counters 1, 2, 3, 4 and 5 onto vacant squares, 4, 5,
6, 3 and 2.

Then pile them on top of counter 5 on square 2.

Move counters 6, 7, 8 and 9 onto vacant squares, 4, 5, 6
and 3.

Then pile them onto counter 9 on square 3.

Next, counters 10, 11 and 12 on vacant squares 4, 5 and 6.

Now pile them on counter 12 on square 4.

Then move counters 13 and 14 onto vacant squares 6 and 5.

And now pile counter 13 on counter 14 on square 5.

Move counter 15 onto square 6, which is left vacant.

Place counter 13 on square 1, then counters 14 and 13 on
counter 15.

Now place counters 10 and 11 on the vacant squares,
then 10, 11 and 12 on square 6, on top of the others.

Now move counters 6, 7 and 8 onto vacant squares, then
6, 7, 8 and 9 onto square 6.

Finally put 1, 2, 3 and 4 on vacant squares, then all
remaining counters on square 6 to complete the pile.

All done!

Bibliography

Almost all of the following books have had their place on my shelves since around the time I came out of the RAF and took up recreational maths as a hobby, aged around 21. Why did I choose that hobby? Well that's the way my mind was working. I think hobbies tend to choose you.

That there was even the remotest possibility that this hobby would result in a career as enjoyable and multi-faceted as mine has been never even entered my head.

Columbus' Egg by Edi Lanners (Paddington Press, 1978)
Cyclopedia of Puzzles by Sam Lloyd 1914 (Pinnacle Books, 1976)
Figuring; The Joy of Numbers by Shakuntala Devi (Coronet, 1977)
Further Mathematical Puzzles and Diversions by Martin Gardner (Pelican, 1969)
Gentle Art of Mathematics by Dan Pedoe (Pelican, 1958)
Johnny Ball's Think Box by Johnny Ball (Puffin, 1982)
Johnny Ball's Second Thinks by Johnny Ball (Puffin, 1987)
Mathematician's Delight by W.W. Sawyer (Penguin, 1943)
Mathematics, A Human Endeavour by Harold R. Jacobs (W.H. Freeman, 1970)
More Mathematical Puzzles and Diversions by Martin Gardner (Pelican, 1961)
More Puzzles and Curious Problems by Henry Ernest Dudeney, Ed. Martin Gardner (Fontana, 1967)
The Moscow Puzzles by Boris A. Kordemski, Ed. Martin Gardner (Pelican 1972)

Penguin Book of Curious and Interesting Puzzles by David Wells (Penguin, 1992)

Penguin Book of Curious and Interesting Geometry by David Wells (Penguin, 1991)

Prelude to Mathematics by W.W. Sawyer (Pelican, 1955)

Riddles in Mathematics by Eugene P. Northrop (Pelican, 1944)

The Geometry by Rene Descartes (1673)

Think of a Number by Johnny Ball (BBC, 1979)

The World of Mathematics by James R. Newman (George Allen and Unwin, 1956)

… and many, many more over the past 50 years.